David Copperfield

塊肉餘生記

商務印書館

Name of Book: David Copperfield
Author: Charles Dickens
Retold by: Derek Sellen
Editor: Rebecca Raynes
Design and art direction: Nadia Maestri
Computer graphics: Sara Blasigh
Illustrations: Simone Massoni
Picture research: Laura Lagomarsino
Picture Credits: Victoria and Albert Museum, London, UK/Bridgeman Art Library: 5; The
Granger Collection, New York: 36-7; Ipswich Borough Council Museums and Galleries,
Suffolk, UK/Bridgeman Art Library: 63
Edition: © 2004 Black Cat Publishing
                an imprint of Cideb Editrice, Genoa, Canterbury

系 列 名：Black Cat 優質英語階梯閱讀 · Level 2
書　　名：塊肉餘生記
責任編輯：黃家麗
封面設計：張　毅
出　　版：商務印書館 (香港) 有限公司
　　　　　香港筲箕灣耀興道 3 號東滙廣場 8 樓
　　　　　http://www.commercialpress.com.hk
發　　行：香港聯合書刊物流有限公司
　　　　　香港新界大埔汀麗路 36 號中華商務印刷大廈 3 字樓
印　　刷：中華商務彩色印刷有限公司
　　　　　香港新界大埔汀麗路 36 號中華商務印刷大廈
版　　次：2015 年 3 月第 4 次印刷
　　　　　©2004 商務印書館 (香港) 有限公司
　　　　　ISBN 978 962 07 1704 8
　　　　　Printed in Hong Kong

出版説明

　　本館一向倡導優質閱讀，近年來連續推出了以"Q"為標識的 "Quality English Learning 優質英語學習"系列，其中《讀名著學英語》叢書，更是香港書展入選好書，讀者反響令人鼓舞。推動社會閱讀風氣，推動英語經典閱讀，藉閱讀拓廣世界視野，提高英語水平，已經成為一種潮流。

　　然良好閱讀習慣的養成非一日之功，大多數初、中級程度的讀者，常視直接閱讀厚重的原著為畏途。如何給年輕的讀者提供切實的指引和幫助，如何既提供優質的學習素材，又提供名師的教學方法，是當下社會關注的重要問題。針對這種情況，本館特別延請香港名校名師，根據多年豐富的教學經驗，精選海外適合初、中級英語程度讀者的優質經典讀物，有系統地出版了這套叢書，名為《Black Cat 優質英語階梯閱讀》。

　　《Black Cat 優質英語階梯閱讀》體現了香港名校名師堅持經典學習的教學理念，以及多年行之有效的學習方法。既有經過改寫和縮寫的經典名著，又有富創意的現代作品；既有精心設計的聽、説、讀、寫綜合練習，又有豐富的歷史文化知識；既有彩色插圖、繪圖和照片，又有英美專業演員朗讀作品的 CD。適合口味不同的讀者享受閱讀之樂，欣賞經典之美。

　　《Black Cat 優質英語階梯閱讀》由淺入深，逐階提升，好像參與一個尋寶遊戲，入門並不難，但要真正尋得寶藏，需要投入，更需要堅持。只有置身其中的人，才能體味純正英語的魅力，領略得到真寶的快樂。當英語閱讀成為自己生活的一部分，英語水平的提高自然水到渠成。

商務印書館 (香港) 有限公司
編輯部

使用說明

① 應該怎樣選書？

按閱讀興趣選書

《Black Cat 優質英語階梯閱讀》精選世界經典作品，也包括富於創意的現代作品；既有膾炙人口的小說、戲劇，又有非小說類的文化知識讀物，品種豐富，內容多樣，適合口味不同的讀者挑選自己感興趣的書，享受閱讀的樂趣。

按英語程度選書

《Black Cat 優質英語階梯閱讀》現設 Level 1 至 Level 6，由淺入深，涵蓋初、中級英語程度。讀物分級採用了國際上通用的劃分標準，主要以詞彙（vocabulary）和結構（structures）劃分。

Level 1 至 Level 3 出現的詞彙較淺顯，相對深的核心詞彙均配上中文解釋，節省讀者查找詞典的時間，以專心理解正文內容。在註釋的幫助下，讀者若能流暢地閱讀正文內容，就不用擔心這一本書程度過深。

Level 1 至 Level 3 出現的動詞時態形式和句子結構比較簡單。動詞時態形式以現在時（present simple）、現在時進行式（present continuous）、過去時（past simple）為主，句子結構大部分是簡單句（simple sentences）。此外，還包括比較級和最高級（comparative and superlative forms）、可數和不可數名詞（countable and uncountable nouns）以及冠詞（articles）等語法知識點。

Level 4 至 Level 6 出現的動詞時態形式，以現在完成時（present perfect）、現在完成時進行式（present perfect continuous）、過去完成時（past perfect continuous）為主，句子結構大部分是複合句（compound sentences）、條件從句（1st and 2nd conditional sentences）等。此外，還包括情態動詞（modal verbs）、被動形式（passive forms）、動名詞（gerunds）、

短語動詞（phrasal verbs）等語法知識點。

　　根據上述的語法範圍，讀者可按自己實際的英語水平，如詞彙量、語法知識、理解能力、閱讀能力等自主選擇，不再受制於學校年級劃分或學歷高低的約束，完全根據個人需要選擇合適的讀物。

② 怎樣提高閱讀效果？

　　閱讀的方法主要有兩種：一是泛讀，二是精讀。兩者各有功能，適當地結合使用，相輔相成，有事半功倍之效。

　　泛讀，指閱讀大量適合自己程度（可稍淺，但不能過深）、不同內容、風格、體裁的讀物，但求明白內容大意，不用花費太多時間鑽研細節，主要作用是多接觸英語，減輕對它的生疏感，鞏固以前所學過的英語，讓腦子在潛意識中吸收詞彙用法、語法結構等。

　　精讀，指小心認真地閱讀內容精彩、組織有條理、遣詞造句又正確的作品，着重點在於理解"準確"及"深入"，欣賞其精彩獨到之處。精讀時，可充分利用書中精心設計的練習，學習掌握有用的英語詞彙和語法知識。精讀後，可再花十分鐘朗讀其中一小段有趣的文字，邊唸邊細心領會文字的結構和意思。

　　《Black Cat 優質英語階梯閱讀》中的作品均值得精讀，如時間有限，不妨嘗試每兩個星期泛讀一本，輔以每星期挑選書中一章精彩的文字精讀。要學好英語，持之以恆地泛讀和精讀英文是最有效的方法。

③ 本系列的練習與測試有何功能？

　　《Black Cat 優質英語階梯閱讀》特別注重練習的設計，為讀者考慮周到，切合實用需求，學習功能強。每章後均配有訓練聽、說、讀、寫四項技能的練習，分量、難度恰到好處。

聽力練習分兩類，一是重聽故事回答問題，二是聆聽主角對話、書信朗讀、或模擬記者訪問後寫出答案，旨在以生活化的練習形式逐步提高聽力。每本書均配有 CD 提供作品朗讀，朗讀者都是專業演員，英國作品由英國演員錄音，美國作品由美國演員錄音，務求增加聆聽的真實感和感染力。多聆聽英式和美式英語兩種發音，可讓讀者熟悉二者的差異，逐漸培養分辨英美發音的能力，提高聆聽理解的準確度。此外，模仿錄音朗讀故事或模仿主人翁在戲劇中的對白，都是訓練口語能力的好方法。

閱讀理解練習形式多樣化，有縱橫字謎、配對、填空、字句重組等等，注重訓練讀者的理解、推敲和聯想等多種閱讀技能。

寫作練習尤具新意，教讀者使用網式圖示（spidergrams）記錄重點，採用問答、書信、電報、記者採訪等多樣化形式，鼓勵讀者動手寫作。

書後更設有升級測試（Exit Test）及答案，供讀者檢查學習效果。充分利用書中的練習和測試，可全面提升聽、説、讀、寫四項技能。

❹ 本系列還能提供甚麼幫助？

《Black Cat 優質英語階梯閱讀》提倡豐富多元的現代閱讀，巧用書中提供的資訊，有助於提升英語理解力，擴闊視野。

每本書都設有專章介紹相關的歷史文化知識，經典名著更有作者生平、社會背景等資訊。書內富有表現力的彩色插圖、繪圖和照片，使閱讀充滿趣味，部分加上如何解讀古典名畫的指導，增長見識。有的書還提供一些與主題相關的網址，比如關於不同國家的節慶源流的網址，讓讀者多利用網上資源增進知識。

Contents

This story is recorded in full. 故事錄音
🎧 This symbol indicates the beginning of a recording and track number. 錄音和音軌標記

Charles Dickens and 'David Copperfield'

Charles Dickens was born in 1812 in Portsmouth. Later, his parents moved to London and to Chatham in Kent. When he was a boy his life was difficult but as an adult Dickens had an international reputation as a great writer.

Dickens called *David Copperfield* 'my favourite child'. He considered it his favourite novel. It is easy to understand this. David's story is similar to Charles Dickens's life. Like David,

Charles Dickens (1859) by William Powell Frith.

Dickens worked as a young boy in a factory, [1] then earned money by reporting the debates in parliament. He later became a popular author. Like David's friend, Mr Micawber, Dickens's father went to prison because he owed [2] money. David's initials, D.C., are the reverse of C.D., Charles Dickens. David is a mirror [3] reflection of the author.

David Copperfield is typical of all his work. Dickens understood children and created many famous child characters. Young David Copperfield is one of these.

Dickens knew and loved London and small towns like Canterbury and Yarmouth, the three places where the main action takes place. He criticises the unjust things in society, such as cruel [4] boarding schools [5] and child labour. [6] There are comic characters in *David Copperfield* like Mr Micawber who always owes money. Above all, Dickens hated hypocrites, people who pretended to be good but were not. One of the novel's characters, Uriah Heep, is possibly the greatest hypocrite in all Dickens's work.

Dickens lived at the time of Queen Victoria, when Britain was a very rich, powerful nation. However, there was a very big difference between the living conditions of the rich and the poor. Dickens in his novels is always on the side of the poor. He creates characters who are good to others like Aunt Betsey, Traddles and Agnes in this story. He believed that it was possible to reform society if people were more compassionate.

---

1. **factory** : 工廠。
2. **owed** : 欠債。
3. **mirror** :
4. **cruel** : 殘酷的。
5. **boarding schools** : 寄宿學校。
6. **labour** : 勞工。

Read the introduction about Charles Dickens and *David Copperfield*. Are sentences 1.–10. 'Right' (A) or 'Wrong' (B)? If there is not enough information to answer 'Right' (A) or 'Wrong' (B), choose 'Doesn't say' (C).

EXAMPLE:

0. Charles Dickens was born in 1821.
A ☐ Right　　**B** ☑ Wrong　　**C** ☐ Doesn't say

1. Dickens was unhappy all his life.
A ☐ Right　　**B** ☐ Wrong　　**C** ☐ Doesn't say

2. Dickens preferred *David Copperfield* to all his other novels.
A ☐ Right　　**B** ☐ Wrong　　**C** ☐ Doesn't say

3. Dickens used his own experience to create the character of David.
A ☐ Right　　**B** ☐ Wrong　　**C** ☐ Doesn't say

4. Dickens, but not his readers, loved the book.
A ☐ Right　　**B** ☐ Wrong　　**C** ☐ Doesn't say

5. Dickens never wrote about foreign countries in his novels.
A ☐ Right　　**B** ☐ Wrong　　**C** ☐ Doesn't say

6. Dickens wrote about Canterbury in other novels.
A ☐ Right　　**B** ☐ Wrong　　**C** ☐ Doesn't say

7. Dickens didn't agree with child labour.
A ☐ Right　　**B** ☐ Wrong　　**C** ☐ Doesn't say

8. Micawber and Uriah Heep are examples of hypocrites.
A ☐ Right　　**B** ☐ Wrong　　**C** ☐ Doesn't say

9. The rich people did not care about the poor in the time of Queen Victoria.
A ☐ Right　　**B** ☐ Wrong　　**C** ☐ Doesn't say

10. Dickens wanted society to remain as it was.
A ☐ Right　　**B** ☐ Wrong　　**C** ☐ Doesn't say

Summary of the Story:

David Copperfield lived happily with his young, pretty mother and their servant, Peggoty. But then his mother married again. His new father, Mr Murdstone, treated David very badly. He sent David to work in a factory in London when he was ten years old. Finally, he ran away to his strange, eccentric [1] aunt who hated boys! What happened to David after that, as he grew into a young man, fell in love and met an old schoolfriend?

Main Characters

David's mother   *David*   *Peggoty*

Miss Murdstone

Mr Murdstone

Mr Peggoty   *Little Emily*   *Ham*

Suffolk

Yarmouth

1. **eccentric**：古怪的。

School

Steerforth

Traddles

Mr Dick

Dover

Aunt Betsey

Canterbury

Mr Wickfield

Agnes

Uriah Heep

London

Mr Micawber

Dora

Mr Spenlow

BEFORE YOU READ

1 Here are some words from the story. Connect the word to the correct picture.

1. ☐ cart
2. ☐ cliff
3. ☐ cane
4. ☐ cravat
5. ☐ daisy
6. ☐ factory
7. ☐ label
8. ☐ bald
9. ☐ devil
10. ☐ fireplace
11. ☐ lifeboat
12. ☐ rope
13. ☐ suitcase

PART **ONE**

My Mother, Peggoty and Me

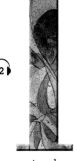

am David Copperfield.

These pages will tell the story of my life.

My father died before I was born and my mother lived alone with a servant, Peggoty. One day, before I was born, one of my father's aunts, Miss Betsey Trotwood, arrived at the house. She was a strange lady.

'Mrs Copperfield,' she said, 'when will the baby girl be born?'

My mother was surprised. 'Maybe it will be a boy,' she said.

'It *must* be a girl,' said Miss Betsey. Betsey Trotwood hated men because of her previous bad marriage.

My mother was very shocked by the aunt. She went upstairs to rest. Soon, the doctor came and went to see her. Aunt Betsey sat downstairs and waited.

A few minutes after midnight, the doctor came downstairs.

'The baby is born. It's a boy,' he said.

Miss Betsey stood up, took her hat, hit the doctor with it, and disappeared from the house. She never came back. And that is the story of the day that I was born.

When I was a young child, I was very happy. I loved my mother and Peggoty. But one year, a man called Mr Murdstone met my mother at church. He often came to the house and he was very kind and friendly to my pretty mother. He had black hair, black eyes, and black clothes. I didn't like him but my mother did.

One day, Peggoty took me to stay with her family in the town of Yarmouth on the coast. I loved Peggoty and I wanted to see the sea. Mr Barkis, the local carrier,[1] took us in his cart.

Peggoty's family lived in an old boat on the beach – it was a very comfortable house. I met Peggoty's brother, Mr Peggoty, a fisherman, and his nephew[2] Ham, and Ham's cousin, Emily. Emily was a beautiful little girl with red hair and blue eyes. I fell in love with her and we spent a lot of time together.

'Would you like to be a lady?' I asked her.

'Yes, I'm afraid of the sea,' she replied. 'I want all my family to be rich, not fishermen.'

I enjoyed my holiday very much but finally the day to go home arrived.

1. **carrier** : 運貨人。
2. **nephew** : 侄子或外甥。

'Goodbye, Master[1] Davy,' said Mr Peggoty. 'Come back soon.'

I rode back in Mr Barkis's cart with Peggoty. When we arrived home, she said, 'David, I must tell you something. You've got a new father.'

I ran into the house. My mother wanted to kiss me but a man in black stopped her. He stood in front of me and shook[2] my hand, then sent me upstairs.

My mother was married to Mr Murdstone.

A few weeks later, his sister, Miss Murdstone, came to live with us. She took the keys of the house from my mother and said, 'Don't worry, Clara, I will organise everything.'

The happy days were finished.

---

1. **Master**：尊稱年齡小，不便稱為先生的男孩。

2. **shook**：握手。

18

UNDERSTANDING THE TEXT

1 **READING**

Did you understand Part One of the story? Are sentences 1.–7. 'Right' (A) or 'Wrong' (B)? If there is not enough information to answer 'Right' (A) or 'Wrong' (B), choose 'Doesn't say' (C).

1. Aunt Betsey wanted the baby to be a boy.
   A ☐ Right   **B** ☐ Wrong   **C** ☐ Doesn't say

2. Mr Murdstone was a tall, strong man.
   A ☐ Right   **B** ☐ Wrong   **C** ☐ Doesn't say

3. Peggoty's brother lived in an old boat.
   A ☐ Right   **B** ☐ Wrong   **C** ☐ Doesn't say

4. Emily was happy that her family were fishermen.
   A ☐ Right   **B** ☐ Wrong   **C** ☐ Doesn't say

5. While David was in Yarmouth, his mother married Mr Murdstone.
   A ☐ Right   **B** ☐ Wrong   **C** ☐ Doesn't say

6. Miss Murdstone wore black clothes like her brother.
   A ☐ Right   **B** ☐ Wrong   **C** ☐ Doesn't say

7. Peggoty didn't like the Murdstones.
   A ☐ Right   **B** ☐ Wrong   **C** ☐ Doesn't say

2 **VOCABULARY**

Read the descriptions of people (1.–5.). What is the word for each one? The first letter is already there. There is one space for each other letter in the word. Complete the words.

0. I look after sick people or children.   n u r s e

1. I am a man who catches fish.   f _ _ _ _ _ _ _ _

2. I am the son of your brother or sister.   n _ _ _ _ _

3. I am the brother of your father or mother.   u _ _ _ _

4. I am a boy or man who is the main character in a story.   h _ _ _

5. I clean, cook or do other work in the home for money.   s _ _ _ _ _ _

LISTENING

You will hear some information about Mr Barkis. Listen and complete questions 1–10.

MR BARKIS

JOB: the local (**1**)...................................... .

He carried (**2**).................................... and (**3**)....................................
from place to place.

For his job, he used a (**4**).................................... and cart.

He did not (**5**).................................... his money but
(**6**).................................... it.

He asked questions about (**7**).................................... .

David told him that she (**8**).................................... well and
(**9**).................................... the house.

Mr Barkis sent her a (**10**).................................... .

4 **WRITING**

David is writing home from Yarmouth. Add one or two sentences to his letter:

Dear Mummy,

I am enjoying my holiday. I am staying with Mr Peggoty and his family. They live in an old boat but it is very comfortable. ....
....
....
....

I will come home soon.

love,

David. X X X

PART **TWO**

Mr Murdstone

r Murdstone hated me.

'What do I do,' he asked me, 'if I have a dog or a horse that doesn't do what I want?'

'I don't know,' I said.

'I hit him,' said Mr Murdstone.

I understood.

My mother was gentle and kind. But when she wanted to kiss me, Mr Murdstone and his sister said, 'Clara, don't be silly. Be strong!'

Mr Murdstone gave me lessons. He asked difficult questions. I was very nervous and I couldn't answer. 'If you don't study your lessons, we must punish you,' he said.

21

My mother began to protest.

'Be strong, Clara,' said Miss Murdstone.

Her brother took me upstairs and began to hit me with a long cane. I turned my head. I bit ¹ his hand. He was very angry and hit me very hard. Then he closed the door and locked it. That night, Peggoty came to the door and said, 'Master Davy, I love you. I'll look after you and your mother.'

Finally, Mr Murdstone unlocked the door. He sent me to a school near London. When I arrived, the headmaster, ² Mr Creakle, put a notice on my back. It said: 'Be careful – he bites.'

It was a horrible school. Fortunately, I made two friends. One was a boy called Traddles, the other was Steerforth. He was a good-looking older boy and came from a rich family. He called me 'Daisy' and protected me from the other boys and the teachers. Mr Creakle was also afraid to hit Steerforth.

'Give me your pocket money, ³ Daisy,' said Steerforth, 'and I'll buy food and drink for everyone. The boys will all like you then.'

He was my hero. When Mr Peggoty and Ham visited me, I introduced them to my new friend and protector. ⁴

When I returned home for the summer, my mother had a new baby. I wanted to hold my little brother but the Murdstones sent me upstairs. I was very sad and was happy to return to Mr Creakle's school.

On my birthday, Mr Creakle called me to his room.

'Was your mother well when you left her?' his wife asked.

I didn't answer. I knew there was a problem.

1. **bit** : 咬。
2. **headmaster** : 校長。
3. **pocket money** : 零用錢。
4. **protector** : 保護者。

'She's ill,' Mrs Creakle said.

I began to cry.

'She's very ill,' she said.

Now I understood.

'She's dead.'

I cried and cried. I went home immediately. Peggoty told me about my mother's death. And she told me how the baby also died. After the funeral, [1] Peggoty sat with me in my room and comforted me. Miss Murdstone said, 'We don't need a servant,' and Peggoty left. But first, she took me away to see her family in the boat-house. It was good to see Mr Peggoty and Ham and Little Emily.

Peggoty decided to marry Mr Barkis. 'I will always love you, Master Davy,' she promised. 'I'll always be here to help you.'

When I returned home, Mr Murdstone didn't speak to me for a long time. Then, one day, he said, 'Education is expensive, David. You must go to London and learn to be independent [2] in the world!'

---

1. **funeral** : 葬禮。
2. **independent** : 自力更生。

UNDERSTANDING THE TEXT

 READING

Did you understand Part Two of the story?
Answer these questions with a phrase or a short sentence.

0. Did Mr Murdstone threaten David? ...*Yes, he did.*...........

1. Why did Mr Murdstone punish David?

2. How did David defend himself?

3. Who talked to David when he was in a locked room?

4. What did Mr Creakle write on the notice?

5. What were the names of David's two school friends?

6. When David went home, what was different about his family?

7. What did Mrs Creakle tell David about his mother?

8. Where did Mr Murdstone plan to send David?

2 **GRAMMAR**

THE PAST SIMPLE FORM OF VERBS

Complete these sentences by using verbs from the box below in the
Past Simple form.

	be	bite	buy	cry	die	
forget	give	hear	hit	put	send	take

1. Mr Murdstone .... David lessons but David always
.... the right answers.

2. David .... Mr Murdstone, who then .... David
severely.

3. He .... David to a school near London where they ....
a notice on his back.

4. Steerforth .... David's money and .... food and drink
for the boys.

5. David's mother .... while he .... away at school.

6. David .... when he .... the sad news.

3 WRITING

Is your school better than David's school? Write five sentences about your school.

For example:

0. My English teacher is very kind and clever. ....

00. There are twenty students in my class. ....

1. ....

2. ....

3. ....

4. ....

5. ....

Now write five things that you want to change in your school.

0. I want to have more time for sports. ....

1. ....

2. ....

3. ....

4. ....

5. ....

T: GRADE 5

4 SPEAKING

Topic – School

Find a picture of a classroom. Use the following questions to help you talk about it.

1. Is it a picture of a typical classroom?

2. Is it similar to your classroom? Why?/Why not?

3. Is a lesson being taught at the moment? If so, can you see what subject?

4. What is your favourite subject at school? Why?

5. How long have you studied this subject?

6. Which subjects do you think you will study in the future?

PART **THREE**

Aunt Betsey

n London, I worked in a factory. I washed and put labels on bottles all day. It was very hard work and the other boys laughed at me and treated me badly. I was ten years old with no father or mother. I was very unhappy.

I lived with Mr and Mrs Micawber and their children. Mr Micawber was a tall man with a bald head [1] like an egg. He always spent more money than he had. He always owed money to shopkeepers. [2]

'Mr Micawber will be a great man, one day,' Mrs Micawber

1. **bald head** : 禿頭。
2. **shopkeepers** : 店主。

said. 'He is the father of my children and my husband. I will never leave him!'

Mr Micawber went to prison because he owed so much money. Finally, the Micawbers left London to go to Plymouth. Now I had no friends. I decided to run away. I knew that my father's aunt, Betsey Trotwood, lived near Dover, about eighty miles [1] from London. I didn't know her but she was my only relative [2] in all the world.

I had a little money and a suitcase but a young man in the street stole [3] everything from me. I decided to walk to Dover! The journey took a few days. I sold some of my clothes to buy food. There were bad people on the road. I was very afraid. I finally arrived in Dover. I was dirty, penniless, [4] weak [5] and very hungry.

In Dover, I asked where Miss Betsey Trotwood lived. I walked up to the top of a cliff. There I found a small cottage.

When she saw me, Miss Betsey shouted, 'No boys! No boys here!'

'I am David Copperfield.'

She was very surprised. She called Mr Dick, a man who stayed with her; he was simple but very gentle and kind. 'What can we do with the boy, Mr Dick?' she asked.

'Give him some food, wash him and put him to bed,' he replied.

---

1. **miles**：英里（1 英里等於 1.6 公里）。
2. **relative**：親人。
3. **stole**：（steal-stole）偷。
4. **penniless**：身無分文。
5. **weak**：身體虛弱。

'Excellent advice, [1] Mr Dick,' she said.

I soon discovered that Aunt Betsey was very kind. She wrote to Mr Murdstone and told him where I was. Two days later, he arrived with Miss Murdstone.

'He is a very bad boy, Miss Trotwood. We will take him home and punish him.'

'My brother was very kind to him and found him a job but he ran away,' explained Miss Murdstone. I trembled. [2]

Aunt Betsey asked Mr Dick, 'What can we do with the boy?'

'Buy him some new clothes,' said Mr Dick.

'Leave my house,' said Aunt Betsey to the Murdstones. 'I believe you mistreated [3] this boy!' she shouted. 'I'll look after him.'

I was safe!

1. **advice** : 建議。
2. **trembled** : 發抖。
3. **mistreated** : 虐待。

UNDERSTANDING THE TEXT

1 READING

Did you understand Part Three of the story?
Complete these sentences with a word or phrase from the text.

0. In London, David had to do very ....hard.......... work.

1. Mr Micawber always owed .... to shopkeepers.

2. .... said: 'My husband will be a great man!'

3. After the Micawbers went to Plymouth, David decided to
....

4. Aunt Betsey was his .... in all the world.

5. He was often afraid because there were .... on the
road.

6. Aunt Betsey lived in a .... near Dover.

7. She said that Mr Dick's advice was ....

8. Aunt Betsey didn't .... the Murdstones' lies.

2 LISTENING

Listen to a conversation about the names which other characters use
for David. Which name does each person use for David? Write a
letter A–H next to each person.

0. Mr Murdstone	A	**A**	David Copperfield
1. Peggoty	☐	**B**	Trotwood
2. Steerforth	☐	**C**	Betsey
3. Agnes	☐	**D**	Daisy
4. Aunt Betsey	☐	**E**	Doady
5. Dora	☐	**F**	Trot
		G	Davy
		H	Master Davy

3 VOCABULARY
PARTS OF THE LANDSCAPE

Aunt Betsey lives near a cliff. Can you match these words with the pictures A–I? Use a dictionary if necessary.

0. a cliff ......A......   **5.** a stream ....

1. a mountain ....   **6.** a valley ....

2. a hill ....   **7.** a bay ....

3. a lake ....   **8.** an island ....

4. a bridge ....

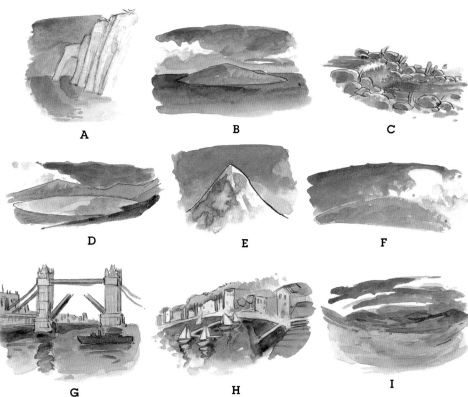

A   B   C

D   E   F

G   H   I

Do you know the names of any other parts of a landscape[1]?

Write them here: ....

Draw pictures of them if you like.

---

1. landscape : 風景。

4 GRAMMAR

LET'S

Read what Mr Dick said (1.–5.). Write the sentences again, using *Let's...*

0. I suggest that we give David some pocket money.
   Let's give David some pocket money...

1. Shall we buy him some new clothes? ....

2. Why not wash him and put him to bed? ....

3. How about flying my kite on the cliffs? ....

4. I think that it's a good idea for us to call him 'Trotwood'.
   ....

5. Why don't we go to Canterbury? ....

5 WRITING

Read these two notes about the route from London to Dover.
For questions 1–5, write the information on the route summary.

ROUTE FROM LONDON TO DOVER:
Take the road from London to Faversham. On the way you'll pass through Chatham where Dickens lived. He later lived in Rochester as an adult. It's on the River Medway – you'll go through it just before Chatham.
At Faversham, follow the main road straight to Dover.

12th March
Sue, I forgot to tell you a few things about the route. If the Dover road is still closed, you'll have to go through Folkestone after you leave Faversham. And on the original route, don't miss stopping off at Canterbury (just after Faversham) to see the house of Agnes – it's an old inn that Dickens describes in *David Copperfield*!
love Jill

ROUTE FROM LONDON TO DOVER
PLACES ON THE ROUTE:

First, pass through [1] [   ]
Then [2] [   ]
Then [3] [   ]
Then [4] [   ]

ALTERNATIVE ROUTE:

Pass through [5] [   ]

PART FOUR

Mr Wickfield, Agnes and Uriah

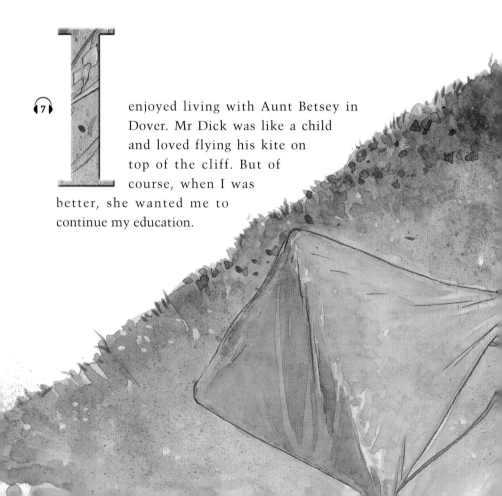

I enjoyed living with Aunt Betsey in Dover. Mr Dick was like a child and loved flying his kite on top of the cliff. But of course, when I was better, she wanted me to continue my education.

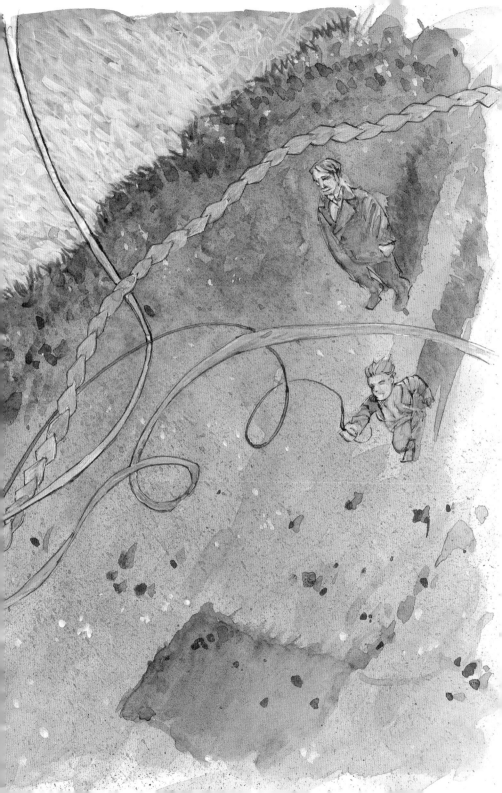

We went to the old cathedral city [1] of Canterbury, near Dover, and she took me to the house of Mr Wickfield. When we arrived, a strange young man came to hold Miss Betsey's horse. His name was Uriah Heep, Mr Wickfield's clerk. [2] He was tall and thin and moved his body like a snake.

Mr Wickfield was Aunt Betsey's lawyer [3] and was responsible for her money.

'Would you like to live with Mr Wickfield and his daughter Agnes and go to school near the cathedral, with Doctor Strong?' asked my aunt.

'Yes,' I replied immediately.

The next few years were very happy. Agnes was always calm and always gave me good advice. She was like a sister to me. She loved her father very much and did everything to help him. His wife was dead and he was very sad without her. My teacher, Doctor Strong, was a generous, intelligent man and I learnt a lot at his school. I fell in love with Canterbury girls many times as I grew older, had fights with local Canterbury boys and did all the things that boys usually do.

I didn't like Uriah Heep. As the years passed, he began to have a strange power over Mr Wickfield.

'Maybe, one day, you will be Mr Wickfield's partner,' I said to Uriah. He moved his body like a snake.

'No, no. You're very kind, Mr Copperfield. I am much too umble. [4] But please come to visit me and my mother in our umble home.'

---

1. **cathedral city**：有大教堂的城市。
2. **clerk**：文員。
3. **lawyer**：律師。
4. **umble**：烏利亞把 humble 讀成 umble，此處解作卑微。

I met Mrs Heep. She was exactly like her son.

'We are very umble, Mr Copperfield,' she said.

At this time, the Micawbers came to Canterbury. They were very surprised to hear my story. They met Uriah but soon left Canterbury again because they owed money.

Finally it was time to leave school. Aunt Betsey asked me to decide on a profession. [1] 'Visit Peggoty in Yarmouth,' she suggested. 'While you are there, think carefully about your future. Then we will decide.'

I was a young man now. I travelled to London alone and stayed there before I continued to Yarmouth. At the inn, [2] I was surprised to meet Steerforth. At first he didn't know me.

'Ah, it's Daisy,' he said.

I was very happy to see him. I invited him to come to Yarmouth with me.

'I'm studying at Oxford but I have a few free days,' he said. 'I'd like to meet these fishermen.'

I met Steerforth's mother, a very arrogant [3] lady, and her companion, Rosa Dartle. Then we left for Yarmouth.

1. **profession** : 專業。
2. **inn** : 旅館。
3. **arrogant** : 傲慢的。

UNDERSTANDING THE TEXT

1 **READING**

Did you understand Part Four of the story?
Answer these questions with a phrase or a short sentence.

0. What did Mr Dick play with on the top of the cliff? ..A kite...........

1. Where did Mr Wickfield and Agnes live?

2. What was Uriah Heep's job?

3. What was the name of David's new teacher?

4. Was David happy in Canterbury?

5. How did the relationship between Uriah and Mr Wickfield change?

6. Who visited Canterbury for a short time?

7. Where did David meet Steerforth again?

8. Why did Steerforth want to go to Yarmouth with David?

2 **READING**

Complete the conversation. What does David say to Doctor Strong?
For questions 1.–5., write the correct letter A–H.

Strong: Did you enjoy school in London?

David: **0.** ......A.............

Strong: Were the lessons difficult?

David: **1.** ....

Strong: Were the teachers unkind?

David: **2.** ....

Strong: Well, I hope you enjoy your
lessons in Canterbury.

David: **3.** ....

Strong: Where are you living?

David: **4.** ....

Strong: He is my old friend. Do you like it
there?

David: **5.** ....

A No, I didn't.

B So do I, sir.

C In Mr Wickfield house.

D No, they were easy.

E Yes, I like.

F I'm afraid so.

G Yes, very much.

H In Mr Wickfield's house.

3 **GRAMMAR**
'S OR S'?
Rewrite these sentences, using 's or s'.

0. Aunt Betsey is the sister of the father of David.
   Aunt Betsey is David's father's sister.

1. Uriah Heep is the clerk of Mr Wickfield.

   ....

2. The young wife of Doctor Strong is very beautiful.

   ....

3. Mr Dick likes playing games for children.

   ....

4. Doctor Strong has a school for boys.

   ....

5. The wages of the teachers are high.

   ....

6. What is the name of the daughter of Mr Wickfield?

   ....

4 **WRITING**
In the story, it says: *Uriah moved his body like a snake.*
We sometimes compare people to animals or machines etc. to make
our writing more vivid and exciting. Can you complete these
comparisons? Use the words in the box.

> hippopotamus [1]   lion   monkey   pig
> robot   ~~snake~~   swan

0. He moved his body .... *like a snake* ....
1. The greedy girl ate ....
2. The ballerina [2] moved gracefully ....
3. The boy climbed up the tree ....
4. He roared angrily ....
5. The fat boy lay in the bath ....
6. She obeyed his orders ....

1. **hippopotamus** : 河馬。
2. **ballerina** : 芭蕾舞女主角。

Child Labour

The Industrial Revolution [1] in Britain was a time when the number of factories increased. Many of them used children as workers. They worked in dirty, dark conditions. Often, the children became ill. There

1. **Industrial Revolution**：十八至十九世紀在歐洲展開的工業革命。當時英國變得富有，但全國充斥着貧窮、污染及疾病等許多問題。

were toxic gases in the air around them. There were dangerous machines.

The factory owners paid them little money. They worked for long hours. The children had small fingers and were better at operating some of the machines than adults. They were often very young, even under nine years old. The children could not go to school. So when they were adults, they continued to work in hard jobs for little money. It was a terrible system.

Charles Dickens himself worked in a factory when he was twelve years old. He put the labels on the containers. David Copperfield must do similar work because his stepfather, Mr Murdstone, does not want to pay for his education. A lot of people read Charles Dickens's books about the condition of children in England at this time, such as *Oliver Twist*. This helped to change the situation.

In 1802, a new law protected poor children under nine years old in cotton mills. [1] In 1819, the law covered all children. But many factory owners broke the law. In 1833, Parliament passed a new law. Government inspectors [2] went into factories and stopped the owners from mistreating young children. Other European countries copied the British laws and the situation improved. But even today, in some countries, children must work in terrible conditions. This is very sad.

Girls and women working in a cotton mill in Lancashire in the 19th century.

1. **cotton mills** : 棉織廠。
2. **inspectors** : 督察。

1 Look at the statements below and decide if each one is correct or incorrect. If it is correct, mark A. If it is not correct, mark B.

		A	B
1.	Children enjoyed working in the factories.	☐	☐
2.	It was dangerous for children to work in the factories.	☐	☐
3.	Child labour was cheap and effective for the factory owners.	☐	☐
4.	Dickens worked in a factory as a child.	☐	☐
5.	His novels didn't affect his readers' opinions about child labour.	☐	☐
6.	The new laws of 1802 and 1819 helped to protect children in factories.	☐	☐
7.	Dickens helped to write the law of 1833.	☐	☐
8.	Child labour exists nowhere in the world today.	☐	☐

T: GRADE 5

 2 **Topic – Jobs**
Find a picture of a job which is often done by schoolchildren. Now use the following questions to help you talk about it.

1. What is the job?
2. Is it difficult/tiring?
3. Have you ever done this job?
4. Do you know anyone else who has done this job?
5. What other jobs can you think of that children do?

PART **FIVE**

Doctors' Commons

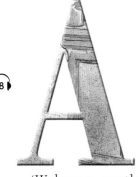

At Yarmouth, I introduced Steerforth to Mr Peggoty, Ham and Little Emily in the boat-house. Little Emily was now a beautiful young woman and Ham was a strong handsome fisherman.

'Welcome, gentlemen,' said Mr Peggoty. 'This is a special day. Ham and Little Emily are engaged.'[1]

Steerforth and I congratulated[2] them. We spent a very nice evening with them.

Afterwards, I said to Steerforth, 'They are a very good-looking couple.'

1. **engaged**：訂婚。
2. **congratulated**：恭賀。

43

'Yes, Daisy. But Ham is too stupid to be a husband for that beautiful girl,' he said.

'You're not serious, Steerforth. I know that you liked them all very much.'

We were in Yarmouth for two weeks. While I stayed with Peggoty and Mr Barkis, Steerforth explored the area. He went out to sea with Mr Peggoty and the other fishermen, bought them drinks and often disappeared alone. He bought a boat and renamed it *The Little Emily*. He gave it to Mr Peggoty. When we left, the fishermen all said goodbye.

In London, my aunt asked me to come to Doctors' Commons. It was an area where lawyers had their offices. She introduced me to Mr Spenlow, an important lawyer. 'I will give Mr Spenlow a thousand pounds,' she said. 'You will work with him and learn about the law. Then one day, you will become a lawyer in Doctors' Commons yourself. Do you like this idea?'

'Thank you, aunt. You are very generous.'

I moved to London and began my new life. One day, I invited Steerforth and his Oxford friends to dinner. We drank a lot of wine. Then we went to see a play at the theatre. I was surprised to see Agnes in the audience. [1] I was very drunk [2] but next day she sent me a kind letter, asking me to visit her while she was in London.

'Steerforth is a bad friend for you,' she told me. 'But I need your help. My father has serious problems. He is afraid of Uriah Heep. He is making him his partner in the law office. I don't understand. My father is very unhappy.'

Uriah and Mr Wickfield were also in London. I invited Uriah to

1. **audience** : 觀眾。
2. **drunk** : 喝醉。

dinner because I wanted more information.

'I am a very umble person,' he said. 'Mr Wickfield has some financial problems. I want to help him. I am his partner now.'

'Yes, Agnes told me.'

'Ah, Agnes. I love her, Mr Copperfield. I am too umble to hope to marry her. But I adore her.'

I hated him when he talked about Agnes like that but I could do nothing. Uriah had the power now.

Agnes returned to Canterbury and I continued working in Doctors' Commons. Then something happened and I forgot everything. Mr Spenlow invited me to dinner at his house. He introduced me to his daughter, Dora. She was small and very pretty.

I fell in love.

UNDERSTANDING THE TEXT

1 READING

Did you understand Part Five of the story? Are sentences 1.–7. 'Right' (A) or 'Wrong' (B)? If there is not enough information to answer 'Right' (A) or 'Wrong' (B), choose 'Doesn't say' (C).

1. David arrived on the evening of Emily and Ham's engagement.
   A ☐ Right **B** ☐ Wrong **C** ☐ Doesn't say

2. Steerforth said that Ham was not a good husband for Emily.
   A ☐ Right **B** ☐ Wrong **C** ☐ Doesn't say

3. Steerforth gave the boat to Emily.
   A ☐ Right **B** ☐ Wrong **C** ☐ Doesn't say

4. Mr Spenlow paid David a thousand pounds to work for him.
   A ☐ Right **B** ☐ Wrong **C** ☐ Doesn't say

5. Steerforth encouraged David to get drunk.
   A ☐ Right **B** ☐ Wrong **C** ☐ Doesn't say

6. Uriah and Mr Wickfield are partners now.
   A ☐ Right **B** ☐ Wrong **C** ☐ Doesn't say

7. Dora is blonde with blue eyes.
   A ☐ Right **B** ☐ Wrong **C** ☐ Doesn't say

2 LISTENING

You will hear Aunt Betsey talking to Mr Spenlow. Listen and complete questions (1–5).

DAVID COPPERFIELD IN LONDON:
Place of work: (**1**) ....
Accommodation with Mrs (**2**)...
MR SPENLOW:
House in (**3**) ....
Age of daughter: (**4**)..
AUNT BETSEY:
Arrival time in Dover: (**5**)..

3 READING

Read the sentences (1.–5.) below. Choose the best word (A, B or C) for each space.

0. Steerforth ....A......... Ham.
  (A) criticised   **B** took   **C** stood

1. My aunt .... me to Mr Spenlow.
  A greeted   **B** introduced   **C** met

2. I .... Steerforth to dinner.
  A invited   **B** invitation   **C** accepted

3. I didn't behave .... at the theatre because I was drunk.
  A good   **B** polite   **C** well

4. Agnes sent a ...., asking me to see her.
  A note   **B** paper   **C** document

5. Agnes was .... about her father because he had problems.
  A excited   **B** anxious   **C** worry

4 GRAMMAR

PREPOSITIONS

Fill the gaps in these sentences with the correct preposition. Choose the words from the box.

about	at	from	in	of	on	on	to	with	with

1. David depended .... his aunt for money and education.

2. Ham looked forward .... marrying Emily.

3. Mr Spenlow was good .... making money.

4. David spent his money .... food and drink for his guests.

5. 'I don't agree .... you about Ham and Emily,' said Steerforth.

6. Agnes and David talked .... her father's problems.

7. David fell in love .... Dora.

8. David was interested .... the law.

9. Is Mr Wickfield afraid .... Uriah Heep?

10. Steerforth's ideas are different .... David's.

PART **SIX**

Little Emily

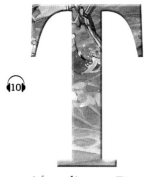

here was someone behind Dora.

'This is my daughter's special friend,' said Mr Spenlow.

It was Miss Murdstone. We shook hands coldly. [1]

After dinner, Dora played the guitar and sang in French. She had a little dog, Jip, that she loved very much. Later, I walked with her in the garden.

'I hate Miss Murdstone,' she told me. 'Papa pays her to look after me but she is very strict. [2] I was in Paris for a year. It was

1. **coldly** : 冷漠地。
2. **strict** : 嚴厲。

48

lovely. You must visit Paris, Mr Copperfield.'

'I will never leave England now that you are here,' I said.

She laughed and kissed her dog. 'Darling[1] Jip!'

During the next weeks, I thought of Dora all the time. I also visited Traddles, my old friend from school, and discovered that he knew Mr and Mrs Micawber. They were back in London. I invited them all to dinner.

After they left, I found a letter from Mr Micawber. 'I have no money,' he wrote. 'My life is a tragedy.'

'Poor Traddles,' I thought. 'Mr Micawber will borrow[2] money from him.' At this moment, Steerforth arrived at my rooms.

'I was in Yarmouth,' he told me. 'Mr Barkis is very ill.'

'Poor Peggoty. I must go to them.'

'Yes, but first come home with me and meet my mother again. It could be the last time we meet, Daisy.'

'What do you mean?'

'Nothing. I am very depressed[3] sometimes. But promise me, Daisy, that you will always remember me as I am now.'

I promised him, but I didn't understand.

After I stayed with his arrogant mother and Rosa Dartle, I travelled to Yarmouth. Mr Barkis was in bed, with his hand on a large box. Later that night Mr Barkis died. In the box, we found his will.[4] He left a lot of money to Peggoty and her family.

After the funeral, I went to see Mr Peggoty. Ham called me outside the door. He looked terrible.

---

1. **darling** : 親愛的。
2. **borrow** : 借。
3. **depressed** : 沮喪。
4. **will** : 遺囑。

'Master Davy, read this letter. It's from Little Emily.'

'When you read this, I will be far away. Perhaps he will make me a lady. I am very sorry.'

At first I did not understand.

'Little Emily is with Steerforth! They disappeared together. They are in France.'

Poor Ham! We told Mr Peggoty. He loved his niece very much and was not angry with her. But he went immediately to the beach and destroyed the boat, *The Little Emily*.

I took Mr Peggoty to see Steerforth's mother.

'Can you help us? Your son must marry my Emily or she will lose her reputation.'[1]

'My son is superior to your niece,' she said. 'They can never marry.'

'She is a very bad girl,' said Rosa Dartle.

It was useless.[2] We left the house.

'Master Davy, I will look for my Emily in all the countries of Europe. It will be a long journey but I must find her. She is my precious child and I love her.'

---

1. **lose her reputation**：（此處指）她因私奔而名譽受損。
2. **useless**：（此處指）毫無用處。

UNDERSTANDING THE TEXT

1 **READING**

Did you understand Part Six of the story? Answer questions 1.–8. with a phrase or a short sentence.

0. How did Mr Spenlow describe Miss Murdstone?
  ~~My daughter's special friend.~~

1. Was Miss Murdstone really Dora's friend? ....

2. Who did David invite to dinner? ....

3. Was Mr Barkis a poor man when he died? ....

4. What did Ham show David? ....

5. What did Mr Peggoty do because he was angry? ....

6. Were Mrs Steerforth and Rosa sorry for Emily? ....

7. What did Mr Peggoty decide to do? ....

8. Was Steerforth an honest friend? ....

2 **LISTENING**
PET

You will hear some information about people in the story. Listen and complete these sentences with a word or phrase from the recording.

1. Steerforth was David's ....

2. Steerforth's first name was ....

3. Rosa Dartle was Steerforth's ....'s companion.

4. Mrs Steerforth's .... was in Highgate.

5. Steerforth went to .... after he left Mr Creakle's school.

6. Littimer was Steerforth's ....

7. Steerforth and Emily ran away to ....,
  .... and Italy.

52

VOCABULARY

Read the descriptions 1.–10. of some things that are related to the sea. What is the word for each one? The first letter is already there. There is one space for each other letter in the word.

0. An area of sand or pebbles next to the sea.   b <u>e</u> <u>a</u> <u>c</u> <u>h</u>

1. A tower with a light to warn ships.   l _ _ _ _ _ _ _ _ _

2. Fishermen use these to catch fish.   n _ _ _

3. A place where boats can shelter.   h _ _ _ _ _ _

4. The English traditionally eat these with fish.   c _ _ _ _

5. The line where the sea seems to meet the sky.   h _ _ _ _ _ _

6. Children sometimes make this from sand.   c _ _ _ _ _

7. A town where many ships arrive and leave.   p _ _ _

8. The sea has many of these.   w _ _ _ _

9. A large area of sea, such as the Pacific.   o _ _ _ _

10. A creature with six legs that lives in the sea.   c _ _ _

PROJECT ON THE WEB

Find out about places in *David Copperfield* on the Internet.
Use a search engine to find:

- information about Yarmouth
- information about Dover
- information about Canterbury

Complete this table as much as you can. Don't worry if you can't fill every empty box – the Internet doesn't always tell you everything!

Place	County and area of England	Important information	Approximate population
Yarmouth			
Dover	Kent south-east		
Canterbury		has a cathedral; a place where pilgrims came	33,000

PART **SEVEN**

Dora

did not hate Steerforth. But I was very sad. However, I soon had a reason to be happy. Mr Spenlow invited me to a picnic to celebrate Dora's birthday.

I bought flowers for Dora in Covent Garden[1] market. I wore a new cravat[2] and new boots. I ordered a box of food. Then I travelled to the picnic place in the country. The sun shone. Dora wore a sky-blue dress. She was not a human being. She was a goddess!

Jip barked[3] at me. He was jealous.

1. **Covent Garden**：昔日倫敦主要的花果市場，現在已成為旅遊景點。
2. **cravat**：男裝領巾。
3. **barked**：吠。

'That horrible Miss Murdstone is not here,' she told me. 'She is at her brother's wedding.[1] He is marrying a rich young woman.'

'Poor girl,' I thought. 'She will suffer like my mother.'

I was unhappy because other young men flirted with[2] Dora at the picnic. But her best friend, Julia, spoke to me privately. 'Dora will stay with me next week. Please, visit her there.'

Dora wanted to see me! Of course I visited her. Soon I held her in my arms and told her I loved her. We became secretly engaged. I visited her at Julia's home and at other times sent her secret love letters. She called me 'Darling Doady' in her letters to me.

At that time, my aunt came to see me with Mr Dick. 'I have bad news,' she said. 'I have no more money. I didn't invest it well. I must live with you here.'

I was very sorry for my aunt and decided to work hard[3] to make more money. I became a secretary for Doctor Strong, my old teacher, who now lived in London. I wrote reports for the newspapers about Parliament. Agnes and Traddles helped me to find these jobs. I continued to study with Mr Spenlow and took the first opportunity to see Dora alone.

'I am poor, Dora. Do you still want to marry me?'

'Of course, Doady. We will be very happy.'

'But you must learn to cook and look after a house.'

'Don't say that, Doady. I cannot do that!'

1. **wedding**：婚禮。
2. **flirted with**：調情。
3. **hard**：（此處指）努力。

Dora

Then Mr Spenlow asked to see me. He was with Miss Murdstone.

She opened her handbag. [1] I saw my love letters to Dora.

'I found these and showed them to Mr Spenlow,' she said with an unfriendly smile.

'Mr Copperfield, it was very bad to write to Dora when I did not know. You must never see my daughter again. Do you understand?' said Mr Spenlow.

'But we are engaged...'

'Nonsense. Dora will marry a rich gentleman, not a poor student like you. Go away.'

It was terrible. But that evening, Mr Spenlow had a heart attack and died. We discovered that he owed a lot of money. Dora went to live with two aunts. She and I were both poor and we could not see each other any more!

---

1. **handbag** : 手袋。

UNDERSTANDING THE TEXT

1 READING

Did you understand Part Seven of the story? Complete these sentences with a word or phrase from the story.

0. David was very ....*sad*........ because of Steerforth.

1. Dora's father invited David to a ....

2. David thought that Dora looked like a ....

3. Jip barked because he was ....

4. Mr Murdstone married a ....

5. Dora and David wrote .... love letters to each other.

6. Aunt Betsey .... all her money because she didn't invest it well.

7. Mr Spenlow didn't want David to .... Dora.

8. Dora's .... invited her to stay with them after Mr Spenlow died.

2 VOCABULARY

Here are some pictures of things which lovers often give or send each other. Write the name of each item (1.–5.).

0. ...*a bunch of red roses*.......

2. ....

1. ....

3. ....

4. ....

5. ....

T 3 READING

Read the summary of the early part of *David Copperfield*.
Choose the best word (A, B, C or D) for each space (1.–8.).

David Copperfield (**0**)..Å.. happy as a child (**1**)....... Mr Murdstone married
his mother. His new father sent him (**2**)....... a horrible school and later,
(**3**)....... his mother died, to a factory in London. David (**4**)....... away and,
luckily, his (**5**)....... aunt protected him. He lived with Agnes and Mr
Wickfield in Canterbury and enjoyed his life (**6**)........ . When it was time to
choose a profession, his aunt (**7**)....... him to think about it for a time. He
decided to visit Peggoty in Yarmouth. He invited Steerforth, (**8**)....... met
him by chance during his journey, to accompany him.

0. (A) was	**B** is	**C** been	**D** were
1. **A** if	**B** until	**C** while	**D** when
2. **A** to	**B** at	**C** on	**D** in
3. **A** while	**B** before	**C** during	**D** after
4. **A** escaped	**B** ran	**C** running	**D** runned
5. **A** fathers	**B** fathers'	**C** father's	**D** father
6. **A** their	**B** they're	**C** there	**D** they
7. **A** let	**B** allowed	**C** lets	**D** made
8. **A** who	**B** which	**C** why	**D** whom

T 4 WRITING

Read this letter from Dora to David.

Darling Doady,
Do you love me? I miss you very much. Miss Murdstone watches me all
the time, so it is difficult to write to you. When can we get married?
Write to me soon, Dora.
PS. Can Jip live with us after we are married?

Write an answer to this letter in about 100 words.

PART **EIGHT**

Uriah Heep

gnes came to see my aunt when she knew about her problems. 'How did you lose your money?' she asked.

'I lost it. Your father was not responsible.'

'Good. Poor papa is so weak these days. He is not a good man of business like he was before. Our home is different. Uriah and his mother live with us. I do not often see my father without them. This is terrible.'

Mr Wickfield and Uriah were also in London and they came to collect [1] Agnes. Mr Wickfield looked old and unhappy. Uriah

---

1. **collect** : 接某人回家。

followed him everywhere.

'Mr Wickfield is very kind to my umble mother and her umble son,' said Uriah. 'And the beautiful Miss Agnes is kind to us also.'

Fortunately, Uriah left on business, so my aunt, Agnes, her father and I had dinner together. We talked about my childhood in Canterbury. Mr Wickfield was happy and smiled. I told Agnes about Dora and she gave me good advice. 'Do nothing secret. Write to the two aunts. Ask permission to visit Dora. I am glad that you are happy, David.' She spoke very calmly and lovingly.

I went to Canterbury to stay with them. When I arrived at the old house, there was a new clerk in the office. It was Mr Micawber!

'My friend Heep offered me work here,' he said. But he looked embarrassed[1] and did not talk to me about his job.

While I was there, Mrs Heep watched Agnes all the time. I was angry and mentioned this to Uriah.

'All is fair in love, Mister Copperfield. My umble mother is taking care of Miss Agnes for me. You are my rival.'[2]

'Agnes is a dear sister to me,' I said, 'but I am engaged to another woman.'

'I am glad.' He squeezed[3] my hand in his damp[4] hand.

'But you will never marry Agnes. She is too good for you, Uriah Heep.'

That night at dinner, Uriah raised his glass. 'I am drinking to Miss Agnes. I am an umble person but I adore her. To be her husband...'

Mr Wickfield suddenly stood up and shouted. 'You will never be her husband!'

---

1. **embarrassed** : 尷尬。
2. **rival** : 敵人。
3. **squeezed** : 揑。
4. **damp** : 濕。

'What's the matter? [1] Be careful!' said Heep.

'Look at my daughter and look at this monster,' said Mr Wickfield. 'I am under his control but I will not give him my darling child.'

I tried to calm him. Finally he sat down and cried quietly.

Uriah also sat down and the dinner continued.

Later, I said goodbye to Agnes before I left to return to London.

---

1. **What's the matter?** : 發生甚麼事？

'Please don't protect your father by marrying Uriah. Your heart and your love are too precious.'

'Don't be afraid, David,' she said and smiled affectionately. [1]

When I left the house, Uriah spoke to me. 'I am friends again with Mr Wickfield. I spoke too early. Do you understand, Mr Copperfield? But I can wait.'

1. **affectionately** : 疼愛地。

UNDERSTANDING THE TEXT

1 READING

Did you understand Part Eight of the story? Complete these sentences about the relationships among the characters by adding phrases A–H.

0. Agnes .A. David.

1. Uriah ..... Agnes.

2. Mrs Heep ..... Uriah, Mr Wickfield and Agnes.

3. Uriah ..... Mr Wickfield.

4. Mr Micawber ..... Uriah.

5. Mr Wickfield ..... Uriah.

6. Agnes ..... Uriah.

7. Uriah ..... David.

A advised

B works for

C lives with

D wants to marry

E talked about his plans to

F doesn't like

G does not want to marry

H has a strange power over

2 LISTENING

You will hear some information about Charles Dickens. Listen and complete questions (1–5).

DICKENS'S EARLY BOOKS 1836-44

1837:	The Pickwick Papers
1837-1839:	Oliver Twist
1838-(1)............:	Nicholas Nickleby
(2)............:	The Old Curiosity Shop
1842:	A visit to (3)....................................... and America.
1842	American (4)...
1844:	Martin Chuzzlewit
1844:	A (5)..

3 READING AND WRITING

Read these two notes about Canterbury. Fill in the information in the Tourist Guide.

Near the Chaucer Hotel, you'll find The Thai House Restaurant and also The River Fish Bar. Near the other main hotel in the centre, the County Hotel, you'll find Zizi's and Mexique. And don't forget that the Thruway Hotel, just outside Canterbury, has a medium priced restaurant.

Tina,

Did you get the info I cut out for you about places to eat? The fish restaurant and Mexique are very pricey, so don't eat there unless you've got a fat wallet. The others are OK in price. But there's a really good place where you can eat for under £5. That's cheap!! It's the Fish Café in the centre. Tell me about it when you get back,

Jim

TOURIST GUIDE

Main hotels in the city centre: *Chaucer Hotel.*

(1) ....

Mid-price restaurants in the centre: (2) ....

(3) ....

Low-price restaurants: (4) ....
....

Other recommended eating place (2 kms from city):
(5) ....
....

4 READING AND WRITING

Complete this note. Write ONE word for each space (1–10).

Today, I visited (Example:..*the*...) house where Charles Dickens lived in London. I am very interested (1).................. Dickens, so I stayed there for (2).................. long time. You can see old furniture, pictures (3).................. manuscripts. I saw the study where he (4).................. some of his novels. I think that (5).................. are other Dickens houses at Rochester, Broadstairs and Portsmouth (6).................. the South of England. I (7).................. really like to visit them too. I know that Shakespeare is your favourite writer, so I expect you want (8).................. go to Stratford-on-Avon. (9).................. you free next weekend? We can go together. It (10).................. be very interesting.

Dickens and Education

Charles Dickens wrote many novels about education. For example, *Nicholas Nickleby*, *David Copperfield* and *Hard Times* all describe schools which treated their pupils badly. Dickens himself went to a private school when he was thirteen. The school was Wellington House. The headmaster, William Jones, was a cruel man. He enjoyed punishing the boys. When Dickens wrote *David Copperfield*, he remembered William Jones and created the character of Mr Creakle.

Lessons in Victorian schools were not very interesting. Children usually learnt the '3 R's'[1] – Reading, Writing and 'Rithmetic. They learnt lists of information. The classrooms were usually dark and very cold in winter. There was little fresh air. The windows were very high. The children could not look out and forget their lessons. In a school like Mr Creakle's, the teachers did not earn a lot of money. The frequent punishments were very strict. Teachers used the cane to hit pupils. They also used other methods. In *David Copperfield*, Mr Creakle hangs a notice on David. It says: 'Be careful – he bites'. Most children were afraid of their teachers.

At the beginning of the nineteenth century, most poor children did not go to school. Their parents wanted them to work and earn money. Some young children went to 'Dame[2] Schools'. In these schools, a woman teacher gave them simple lessons in her home.

1. **the 3 R's**：閱讀、寫作和數學三個基本學科。
2. **Dame**：婦女或淑女。

In *David Copperfield*, Agnes Wickfield starts a school. Dickens himself wanted to start a school for poor children. He did not only write novels but also wrote about education in the newspapers. He wanted to help to make things better. Little by little, the laws changed. After the Education Act of 1870, it was obligatory [1] for all children under 12 to go to school. Schools improved and became free.

The Dame School (1887) by Frederick George Cotman.

1. **obligatory**：強制性的。

Now answer these questions. Are sentences 1.–10. 'Right' (A) or 'Wrong' (B)? If there is not enough information to answer 'Right' (A) or 'Wrong' (B), choose 'Doesn't say' (C).

1. Dickens never wrote about good schools.
   A ☐ Right   B ☐ Wrong   C ☐ Doesn't say

2. In *David Copperfield*, Dickens praised William Jones.
   A ☐ Right   B ☐ Wrong   C ☐ Doesn't say

3. Victorian schools used interesting teaching methods.
   A ☐ Right   B ☐ Wrong   C ☐ Doesn't say

4. The classroom windows were high to give the children fresh air.
   A ☐ Right   B ☐ Wrong   C ☐ Doesn't say

5. Teachers often used the cane.
   A ☐ Right   B ☐ Wrong   C ☐ Doesn't say

6. It wasn't necessary to have a qualification to be a teacher.
   A ☐ Right   B ☐ Wrong   C ☐ Doesn't say

7. Before 1870, no poor children went to school.
   A ☐ Right   B ☐ Wrong   C ☐ Doesn't say

8. 'Dame Schools' were usually successful.
   A ☐ Right   B ☐ Wrong   C ☐ Doesn't say

9. Dickens did not do anything practical to improve education.
   A ☐ Right   B ☐ Wrong   C ☐ Doesn't say

10. 1870 was an important date in the history of British education.
    A ☐ Right   B ☐ Wrong   C ☐ Doesn't say

PART **NINE**

Doctor Strong and Annie

I did what Agnes suggested and wrote to Dora's aunts. They agreed to let me visit Dora regularly. I was happy – we were engaged again.

I continued working at the house of Doctor Strong. I was his secretary. The doctor was over 60 years old but he had a young wife, Annie, a good friend of Agnes. There was a lot of gossip. [1] People said that Annie married him because he was rich. They believed that she really loved her cousin, Jack, a lazy young man. Doctor Strong did not know about this gossip.

---

1. **gossip** : 閒言閒語。

Agnes and Mr Wickfield came to stay with Annie Strong. Agnes visited Dora with me; they liked each other.

Of course, Uriah and Mrs Heep came to London too. They wanted to watch Agnes carefully. When Annie lived in Canterbury, she did not like Uriah. He decided to break their friendship. He went to Doctor Strong's study with Mr Wickfield and told the doctor about the gossip.

'Your wife is in love with her cousin. Everyone knows.'

At that moment I came into the room. He turned to me. 'Mr Copperfield, it's true, isn't it?'

I hesitated.[1] I knew the gossip and almost believed it.

'What do you think, partner?' asked Uriah.

'I hope I am wrong,' said Mr Wickfield, 'but sometimes I think that Annie and Jack...'

Doctor Strong said, 'I am responsible, gentlemen. I am an old man and I fell in love with a young woman. It is not fair. She is beautiful and loyal but it is not good for her to have an old husband. I had a dream that we were happy. But you are helping me to wake up and to see the truth.' He left the room sadly.

'Why did you tell him?' I asked Uriah.

'Because I want to separate Agnes and Mrs Strong. She is dangerous to my plans.'

I was very angry and I hit Uriah Heep on his cheek.

He caught[2] my hand. 'You are always against me, Copperfield. But I am still your umble friend.'

The mark of my hand was on his face.

Next day, Mr Wickfield and Agnes returned to Canterbury.

---

1. **hesitated** : 猶疑。
2. **caught** : 抓住。

Doctor Strong was unwell and very sad. His wife did not know why and became very unhappy. But Mr Dick knew how to solve the problem. He brought Annie to Doctor Strong and they talked.

'Dear husband, why do you stay away from me?'

'I love you, Annie, but I don't want you to have a boring life with an old husband. It is better for you to be with your cousin Jack.'

'But I love you. You are so intelligent, so kind and so generous. My cousin Jack is selfish [1] and lazy. I know that people gossip about us but it isn't true. You are all I want.'

'My dear Annie,' said the Doctor. From that day, he and his wife were happy again.

'Mr Dick is a genius,' [2] said my aunt.

1. **selfish** : 自私。
2. **genius** : 天才。

UNDERSTANDING THE TEXT

1 READING

Look at the statements below and decide if each one is correct or incorrect. If it is correct, mark A. If it is not correct, mark B.

		A	B
1.	Annie and Jack loved each other.	☐	☐
2.	Uriah Heep wanted to separate Agnes and her friend Annie.	☐	☐
3.	David defended Annie against Uriah's gossip.	☐	☐
4.	Doctor Strong was angry with his wife.	☐	☐
5.	David hit Uriah on his cheek.	☐	☐
6.	Probably Uriah was very angry with David but he didn't show it.	☐	☐
7.	Annie loved and respected her husband very much.	☐	☐

2 LISTENING

You will hear five short conversations. There is one question for each conversation. For questions 1.-5., put a tick (✔) under the right answer.

0. When did Dickens marry Catherine?

A ✔   B ☐   C ☐

1. When did he publish *David Copperfield*?

A ☐   B ☐   C ☐

2. What was the name of the character in his first successful novel?

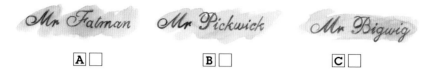

| A ☐ | B ☐ | C ☐ |

3. Which is the second speaker's favourite novel?

| A ☐ | B ☐ | C ☐ |

4. What kind of accident did Dickens experience?

| A ☐ | B ☐ | C ☐ |

5. Where did they bury Dickens?

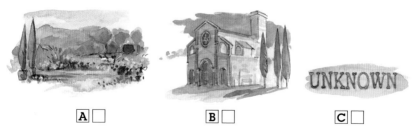

| A ☐ | B ☐ | C ☐ |

3 READING

Complete the conversation. What does Sue say to Tim? Write the correct letter A–H in the gaps (1–5).

Tim: Hello, Sue. How are you?

Sue: (**0**)............H......................

Tim: Would you like to hear some interesting gossip?

Sue: (**1**).................................

Tim: But everyone likes listening to gossip.

Sue: (**2**).................................

Tim: This is about two of the teachers. Mr Griffin and Miss Bellamy are in love.

Sue: (**3**).................................

Tim: Henry told me. He saw them holding hands behind the school. What do you think of that?

Sue: (**4**).................................

Tim: No, other people saw them. Do you want to know any more about their love affair?

Sue: (**5**).................................

A Well, I don't.

B It's probably a lie.

C No, thank you.

D I think so.

E It's not my business.

F Well, I'm not.

G How do you know?

H Fine, thank you.

4 WRITING

Doctor Strong and Annie had problems because people gossiped about them. Write some pieces of gossip about the people in (1.–5.).

Here are two examples:

0. ..My English teacher drinks a bottle of whisky a day.................

00. ..My friend stole some money from his parents.........................

1. The school principal ....

2. The President ....

3. The dentist ....

4. My sister's boyfriend ....

5. The police officer ....

Now complete this sentence:

6. Gossip is bad because ....

PART TEN

Mr Peggoty

D ora and I married. Agnes, Aunt Betsey, Peggoty, Traddles and many other friends were at the wedding. We moved to our small house. I helped Doctor Strong and wrote stories. We were not rich but I had sufficient [1] money to live. But Dora was not good at [2] looking after the house. The servants stole from us or were lazy.

'I cannot learn to cook or do all the other things,' said Dora. 'Dear Doady, I am your child-wife. Think of me like that and love me.'

1. **sufficient** : 足夠。
2. **good at** : 擅長。

My aunt gave me advice. 'Don't ask her to change. Remember Mr Murdstone and your mother. You must love her and be kind.'

Mr Peggoty was back in London. He was grey-haired and tanned[1] after his long journey.

'I looked for Emily and Steerforth in France and Germany and Holland. I cannot find them, Master Davy. But I will continue to look.'

Mrs Steerforth sent for me. Rosa Dartle brought Littimer, Steerforth's personal servant, to talk to me.

'My master and the fisherman's girl were in Italy,' he said coldly. 'He was bored with her and left her. Now he is in Spain. The girl disappeared. That is all I know.'

'You must find the girl,' said Rosa. 'Mrs Steerforth does not want her to trap her son again. Now you can go.'

I told Mr Peggoty this news. Perhaps Emily was in England again.

'There is an old friend of hers in London; Martha. She will tell me if she sees Emily,' he said.

Some months later, Martha contacted me and told me to follow her to a poor area of London. Emily was back in England and was in Martha's room! I sent a message to Mr Peggoty and then went with Martha. When we arrived, we saw a lady go into the room in front of us. It was Miss Dartle. We waited and listened.

'You are a bad girl,' she said cruelly to Emily. 'You stole James Steerforth from his family. He bought you[2] and then rejected[3]

---

1. **tanned**：皮膚被曬黑。
2. **He bought you**：愛米莉被指因收了錢才和斯霍夫私奔。
3. **rejected**：拋棄。

you. You must promise not to see him again.'

'Please be kind to me,' said Emily. 'I will go mad. I loved him and trusted him.'

'You! You loved him!' Rosa Dartle laughed in her face. At that moment, Mr Peggoty arrived. Miss Dartle left. He held Emily.

'Uncle!' she cried and fainted. [1]

'Master Davy,' he said, 'God helped me to find my darling.' He picked her up in his strong arms and carried her downstairs. Little Emily and Mr Peggoty were together again.

Next day, Mr Peggoty told me his plans. 'I will emigrate [2] to Australia with Emily. We will begin a new life away from the gossip in England. Ham knows that she is safe and he forgives her for everything. She is writing to him but she doesn't want to see him. The future will be better for all of us, Master Davy.'

1. **fainted** : 暈倒。
2. **emigrate** : 移居。

UNDERSTANDING THE TEXT

1 READING

Did you understand Part Ten of the story? Here is a summary of the chapter. But there are seven mistakes. There is one example. Can you find and correct six more mistakes?

Nobody came to Dora and David's wedding. Dora was not a good housewife. Aunt Betsey told David to criticise her. At this time, Mr Peggotty returned to England. He asked one of Emily's friends, Mary, to help him. One day, the friend found David and took him to Emily's room. Mrs Steerforth was in the room. She was angry with Emily. Then, Mr Peggotty arrived and found Emily. He was unhappy. He planned to emigrate to Africa with Emily.

0.   A lot of friends came to their wedding.

1.   ....

2.   ....

3.   ....

4.   ....

5.   ....

6.   ....

2 WRITING

Someone is asking you questions about *David Copperfield*. Here are the answers to their questions. Can you write the questions?

0.   Where did David and Steerforth meet? ....?
     At Mr Creakle's school.

1.   ....?
     To Italy and Spain.

2.   ....?
     Steerforth's personal servant.

3.   ....?
     Because she was Emily's friend.

4. ....?

In Martha's room.

5. ....?

Because she was very angry.

6. ....?

He carried Emily downstairs.

7. ....?

Ham was sad but he forgave her.

3 VOCABULARY

Read the definitions (1.–10.) of some verbs. What is the word for each one? The first letter is already there. There is one space for each other letter in the word.

0. Not be able to find something that you previously had.   l o s e

1. Save someone from a dangerous situation.   r _ _ _ _ _

2. Look for somebody or something.   s _ _ _ _ _

3. Disappear, to go out of sight.   v _ _ _ _ _

4. Go to another country to start a new life.   e _ _ _ _ _ _ _

5. Become unconscious through shock, tiredness etc. f _ _ _ _

6. Talk about other people's affairs, spread rumours. g _ _ _ _ _

7. Take things which do not belong to you.   s _ _ _ _

8. Get money for doing a job.   e _ _ _

9. Be responsible for organising a business, household etc.   m _ _ _ _ _

10. Tell somebody about a possible danger, problem etc.   w _ _ _

4 LISTENING

You will hear some information about Charles Dickens. Listen and complete questions (1–5).

THE LATER LIFE OF CHARLES DICKENS

1856:   He moved into (**1**)...............................
(**2**).................:   He met Ellen Ternan, a young
                     (**3**)...............................
1857:   He (**4**)............................... from his wife.
1860s: He read from novels in Britain.
1868:   He gave (**5**)............................... in America.

5 READING AND WRITING

Complete this note. Write ONE word for each space (1–10).

Dear Katie,

How (Example: are.) you? I'm (**1**)..................... holiday at the moment. (**2**)..................... rains every day, so I am in the house. I'm reading a book (**3**)..................... Charles Dickens. The title is 'David Copperfield'. I can't read English very (**4**)....................., so I have got a simplified version. I like the story (**5**)..................... much and I also like the characters. My favourite one (**6**)..................... David's father's aunt, Betsey Trotwood. She often gets angry and behaves strangely (**7**)..................... she has a good heart. When David ran away (**8**)................. his cruel stepfather, she helped him.

I must stop this letter now as it (**9**)..................... getting late. I want to read another chapter before I go (**10**)..................... bed.

Love,
Angelo

PART ELEVEN

Mr Micawber

r Micawber wrote a mysterious letter to me from Canterbury where he still worked as a clerk for Uriah Heep. He organised a meeting in London with me and Traddles.

'Gentlemen,' said Mr Micawber when we met, 'I am in a tragic [1] situation. My employer Heep is a devil. In the beginning I enjoyed working for him but now that I know his true plans, I can't continue to serve him. He is a hypocrite, [2] a serpent, [3] a dishonest man and a liar.'

1. **tragic** : 悲慘。
2. **hypocrite** : 偽君子。
3. **serpent** : 魔鬼。

We took him to see my aunt and Mr Dick. Micawber asked us all to meet him at the offices of Wickfield and Heep in Canterbury in a week's time. 'I will expose [1] Heep!' he promised.

We went to Canterbury on that day and found Mr Micawber at his desk. He took us inside the house. Uriah Heep was surprised to see us. He remembered the last time he met me and the slap [2] on his face.

'Well, I am umbly pleased to see you,' he said. 'Micawber, please return to your desk.'

Mr Micawber did not move. Agnes entered the room at this moment.

'Micawber, leave us!' shouted Uriah. 'Or you will lose your job.'

'There is a hypocrite in the room, sir,' said Mr Micawber, 'and his name is Heep.'

'Go away, all of you,' said Uriah.

'I am working as a lawyer for Mr Wickfield,' said Traddles. 'I am staying. But first I will go and get Mrs Heep.'

He left and returned with Uriah's mother. 'Be umble, Uriah,' she said.

'No. I am the master here. You must all go.'

Mr Micawber made a long speech. 'I know all your secrets, Heep,' he ended. 'Mr Wickfield is afraid of you because you stole money from the business and then persuaded him that he was guilty, [3] not you. You lied and robbed Mr Wickfield's clients. You

---

1. **expose**：揭發。
2. **slap**：巴掌。
3. **guilty**：有罪。

took Miss Betsey Trotwood's money and said that Mr Wickfield was responsible. You are a thief.'

'Prove it, Copperfield and Micawber,' said Uriah.

Mr Micawber produced a notebook. 'This is the evidence. [1] You tried to burn it but I found it in the fireplace. All your secrets are in this little book!'

Traddles took control of the meeting. 'You must give back all the money, Heep. You must tell Mr Wickfield the truth. You must give us your papers.' [2]

'I won't do it!'

'Be umble, Uriah,' said Mrs Heep.

'Do you prefer prison?' asked Traddles.

When he heard this, Uriah agreed to everything. Traddles examined all Uriah's false papers. My aunt received her money and Mr Wickfield was not afraid any more.

'You are wonderful, Mr Micawber,' said my aunt. 'I will give you some money to start a new life in Australia with your family.'

'Mr Micawber will be a great man in Australia,' said his wife.

It was the end of Uriah Heep's power. Agnes and her father were happy together again.

1. **evidence** : 證據。
2. **papers** : 文件。

UNDERSTANDING THE TEXT

1 **READING**

Did you understand Part Eleven of the story? Answer these questions with a phrase or a short sentence.

0. Who made the plan to accuse Uriah Heep?

.Mr.Micawber..

1. Mr Micawber used four words to describe Heep. What are they?

....

2. Who went with David to Canterbury? ....

3. What did Mr Micawber do when Uriah told him to leave?

....

4. What evidence did Mr Micawber have? ....

5. Where did he find the evidence? ....

6. Why did Uriah agree to give back the money? ....

7. How did Aunt Betsey help Mr Micawber to emigrate?

....

8. Were Agnes and her father happy? ....

2 **LISTENING**

You will hear some information about Charles Dickens. Listen and complete questions (1–5).

CHARLES DICKENS

(1)................................., 1870: he died at the age of
(2).................................. .

Probable reason for early death: Overwork.

He wrote novels, **(3)**.................................. and gave readings.

Books about Dickens:

A biography by **(4)**.................................., his best friend.

A modern biography by **(5)**.................................. .

Look at the text in each question. What does it say? Mark the letter next to the correct explanation – A, B, or C.

0.

Governor's Office: interviews by appointment only

A ✓ You must make an appointment if you want to see the Governor.

B ☐ You can only see the Governor in his office.

C ☐ If you want an appointment, you must interview the Governor.

1.

No weapons allowed on prison grounds

A ☐ All weapons must be left on the ground.

B ☐ Do not drop weapons on the ground.

C ☐ You cannot bring a knife into the prison.

2.

RESTRICTED AREA: NO VISITORS BEYOND THIS POINT

A ☐ Only visitors can go into this area.

B ☐ You cannot visit all parts of the prison.

C ☐ There are strict rules in this area.

3.

SECURITY ALERT: Prisoner missing from his cell. All guards to gate B.

A ☐ There is an emergency at gate B.

B ☐ A prisoner has escaped from gate B.

C ☐ Gate B is missing.

4.

SOCCER FIXTURE: prisoners v warders result: 3 – 3

A ☐ There is a football match on the 3rd of March.

B ☐ The prisoners beat the wardens.

C ☐ The prisoners played against the prison staff.

5.

Prison visits take place: weekdays: 9 – 17.30 Sundays: 11 – 15.30 (no visits on Saturdays)

A ☐ Visits finish at 5.30 every day.

B ☐ You cannot visit a prisoner every day of the week.

C ☐ You can only visit the prison on Saturdays.

READING

Read the text below. Choose the best word (A, B, C or D) for each space 1.–8.

THE PRISONERS

Later in his (**0**)....A....., David visited a prison and (**1**)........... many familiar faces. The governor of (**2**).......... prison was Mr Creakle! (**3**).......... were two model prisoners in the cells. Mr Creakle took David to see them. He was (**4**).......... surprised when he recognised Uriah Heep and Mr Littimer, Steerforth's servant. They both said that they were very sorry for (**5**).......... crimes and they hoped that other (**6**).......... could learn from their mistakes. Mr Creakle said that they were the best prisoners (**7**).......... the world. 'They're both hypocrites,' thought David but he said (**8**).......... .

0. (A) life	**B** live	**C** lives	**D** alive
1. **A** seen	**B** sees	**C** saw	**D** see
2. **A** a	**B** the	**C** which	**D** that
3. **A** They	**B** There	**C** Their	**D** They're
4. **A** much	**B** lot	**C** well	**D** very
5. **A** their	**B** his	**C** themselves	**D** there
6. **A** people	**B** person	**C** peoples	**D** persons
7. **A** of	**B** from	**C** to	**D** in
8. **A** anything	**B** something	**C** nothing	**D** everything

5 **WRITING**

Write five sentences which describe Mr Micawber's life in Australia. Use your imagination!

For example:

0. He rode a kangaroo.

00. He learnt how to throw a boomerang.[1]

1. ....

2. ....

3. ....

4. ....

5. ....

1. **boomerang** : 回力鏢。

PART TWELVE

Agnes

ora was ill. She grew weaker every day. I carried her up the stairs every evening but soon she was in bed all the time.

'Darling Doady. Please send Agnes to me.'

Agnes went upstairs while I waited with Jip. Jip was an old dog now. I heard the calm voice of Agnes from Dora's bedroom. Then there was no sound. Finally Agnes came down the stairs and looked silently upwards. I understood. Dora was dead.

I decided to go to Yarmouth to see Peggoty and to give Ham a letter from Emily. She asked him to forgive her. As I travelled, a

terrible storm [1] began. There was a great wind all day and all night.

At Yarmouth, a group of people watched the sea. There was a ship from Spain near the coast. It couldn't enter the harbour [2] and the enormous waves continued to hit it until it began to sink. The lifeboat saved some of the people but there was a man on the ship. He was waving his hands. Nobody could help him.

Suddenly, Ham ran through the crowd. He tied a rope round his body.

'Stop,' said his friends. 'It's too dangerous.'

But Ham was very sad. He didn't care if he lived or died. His friends held the rope and Ham swam towards the ship. He couldn't reach it and came back to the beach. Then he tried again. He disappeared under a huge wave. There was a loud noise and the ship disintegrated, [3] and the man on it disappeared into the sea too.

Later, after the storm, there were two drowned [4] men on the beach. One was Ham. The other was the passenger from the ship. It was Steerforth!

It was a sad time but we continued to make plans. Mr Peggoty, Emily, Mr and Mrs Micawber and their children emigrated to Australia. They went to start a new life and were very successful. My aunt was rich again.

'Why did you say that you lost the money yourself?' I asked. 'Mr Wickfield lost it.'

'I wanted to protect my old friend,' replied my kind aunt.

---

1. **storm** : 風暴。
2. **harbour** : 港口。
3. **disintegrated** : 解體。
4. **drowned** : 淹死。

Agnes

Mr Wickfield was poor now but happy. Agnes gave lessons in the old house. She was a very good teacher and the children loved her.

I was very unhappy and I travelled abroad [1] for two years. I remembered my child-wife, my dead school-friend and poor Ham. Most of all, I thought about Agnes. I understood clearly: I loved her. At last I returned to Aunt Betsey.

'How is Agnes?' I asked.

'She is well,' said my aunt. 'I believe there is a special person in her heart.'

I went to Canterbury to visit her. 'Dear Agnes,' I said, 'my aunt tells me that you are in love. Tell me about the special man.'

Agnes looked away. 'There is no-one. Aunt Betsey is wrong.'

'I love you, Agnes. You are the calm, loving person at the centre of my life. Marry me.'

Agnes smiled. 'Before she died, Dora asked me to take her place. Now you are back and yes, David, I will be your wife.'

'My angel!'

And this is where my story ends.

---

1. **abroad**：到國外。

UNDERSTANDING THE TEXT

1 **READING**

Did you understand Part Twelve of the story? Are sentences 1.–7.
'Right' (A) or 'Wrong' (B)? If there is not enough information to answer
'Right' (A) or 'Wrong' (B), choose 'Doesn't say' (C).

1. Agnes was with Dora when she died.
   A ☐ Right   B ☐ Wrong   C ☐ Doesn't say

2. Steerforth was on the ship that came from Spain.
   A ☐ Right   B ☐ Wrong   C ☐ Doesn't say

3. Ham and Steerforth both drowned.
   A ☐ Right   B ☐ Wrong   C ☐ Doesn't say

4. Agnes taught many different subjects to the children.
   A ☐ Right   B ☐ Wrong   C ☐ Doesn't say

5. David was happy that his wife died.
   A ☐ Right   B ☐ Wrong   C ☐ Doesn't say

6. Probably, Aunt Betsey knew that Agnes loved David.
   A ☐ Right   B ☐ Wrong   C ☐ Doesn't say

7. Agnes and David got married in Canterbury.
   A ☐ Right   B ☐ Wrong   C ☐ Doesn't say

2 **GRAMMAR**
PAST VERBS
Use verbs from the box to fill the gaps. Use the Past Simple tense.

agree	be	begin	blow	break	die	forget
know	know	love	run	sink	think	travel

1. Ham's heart .... when Emily .... away with
   Steerforth.

2. A strong wind .... all day and a lot of ships
   ....

3. .... Ham .... that Steerforth was on the ship?

4. David .... sad because his two friends ....
5. After this, David .... abroad for a long time and
   .... about Agnes.
6. Little by little, he .... the past and .... to look
   forward to the future.
7. Aunt Betsey .... that Agnes .... David.
8. Agnes .... to be David's wife.

PROJECT ON THE WEB

Find out about Dickens and his books on the Internet.
Do you enjoy using the Internet? Can you use it to find out more about
Dickens? Use a search engine to find:

- The names of three novels by Dickens that he wrote apart from *David Copperfield*
- The dates when he originally published these novels
- The ISBN of a paperback edition of each novel. Remember that an ISBN usually has several digits.

Write the information in this table:

title of novel	date	paperback ISBN

EXIT TEST

1 Answer these questions. Are sentences 1.–10. 'Right' (A) or 'Wrong' (B)? If there is not enough information to answer 'Right' (A) or 'Wrong' (B), choose 'Doesn't say' (C).

1. Aunt Betsey is one of the most important people in David's life.
   A ☐ Right   B ☐ Wrong   C ☐ Doesn't say

2. As a child, David had bad experiences in Yarmouth.
   A ☐ Right   B ☐ Wrong   C ☐ Doesn't say

3. After he ran away, David didn't see Mr Murdstone again.
   A ☐ Right   B ☐ Wrong   C ☐ Doesn't say

4. Uriah Heep was born in Canterbury.
   A ☐ Right   B ☐ Wrong   C ☐ Doesn't say

5. Steerforth was not as good as David thought he was.
   A ☐ Right   B ☐ Wrong   C ☐ Doesn't say

6. Mr Peggoty loved his niece Emily very much.
   A ☐ Right   B ☐ Wrong   C ☐ Doesn't say

7. David married Dora because her father was rich.
   A ☐ Right   B ☐ Wrong   C ☐ Doesn't say

8. Mr Micawber discovered Uriah Heep's secrets.
   A ☐ Right   B ☐ Wrong   C ☐ Doesn't say

9. Agnes started a school because she loved teaching.
   A ☐ Right   B ☐ Wrong   C ☐ Doesn't say

10. Ham and Steerforth died on the same day.
    A ☐ Right   B ☐ Wrong   C ☐ Doesn't say

2 Here are descriptions of five characters from the story of *David Copperfield*. Write their names. There is one space for each letter in their names.

0. He is the hero of the story. He learns a lot about life.
D̲a̲v̲i̲d̲   C̲o̲p̲p̲e̲r̲f̲i̲e̲l̲d̲

1. He rises from a clerk to become a partner in a law company.
_ _ _ _ _   _ _ _ _

2. People gossip about her but she is a good, loving wife.
_ _ _ _ _   _ _ _ _ _ _

3. This is the name of a pet. His owner loves him a lot.
_ _ _

4. When he dies, people discover that he had a lot of money.
Mr _ _ _ _ _ _

5. She loves her husband, even though he always has problems with money.
Mrs _ _ _ _ _ _ _ _

3 Complete the information in the table. Write the names of the places.

PLACES IN *DAVID COPPERFIELD*

David spends holidays with his servant's family in:	**(1)**.......................
He works in a factory in:	**(2)**.......................
His aunt and Mr Dick live in:	**(3)**.......................
The Wickfields live in a house in:	**(4)**.......................
He trains to be a lawyer in:	Doctors' **(5)**.......................
Mr Micawber emigrates to:	**(6)**.......................

4 Complete this student book review. Write ONE word for each space (1–8).

DAVID COPPERFIELD by CHARLES DICKENS

I read this book last month and enjoyed it (Example: ....**very**.....)
much. I enjoyed reading about some **(1)**................. the characters,
(2)................. example Uriah Heep. He is a really bad character
but it is difficult **(3)**................. forget him. I thought that Mr
Micawber was very funny. He owes money **(4)**................. everybody
and never pays it back. But at the end **(5)**................. the story, he
helps them to catch Uriah. I liked Dora but she was **(6)**................. a
very good wife. I felt really happy when David asked Agnes to
(7)................. him. She is definitely the best wife for him. I like books
(8)................. a happy ending.

5 Complete the information in the table. Write the names of the characters.

Characters in *David Copperfield*

David's servant: (1) ....

David's stepfather: (2) ....

David's first headmaster: (3) ....

David's father's aunt: (4) ....

David's 'sister': (5) ....

David's teacher in Canterbury: (6) ....

David's school friends: (7) ....

(8) ....

David's employer in London: (9) ....

David's first wife: (10) ....

6 Answer these questions about Charles Dickens and *David Copperfield*.

0. Where was Charles Dickens born? *Portsmouth.*

1. When was he born? ....

2. What did Dickens call *David Copperfield*? ....

3. Where did Dickens work as a young boy? ....

4. Why did Dickens's father go to prison? ....

5. What are the three places where the story takes place? ....

6. Which person in the novel is a famous comic character? ....

7. Which character is a hypocrite? ....

8. Who was the queen during the lifetime of Dickens? ....

9. Was Britain weak and poor at that time? ....

10. Which section of society does Dickens defend in his novels? ......

7 Read the summary of the story below and choose the correct answer for each space 1.-12. Write the correct letter, A, B or C, in each space.

David Copperfield was a (0) $\overset{A}{.}$ child until his mother married Mr Murdstone. His new 'father' was (1) ... to David but luckily his nurse, Peggoty, helped David. David had a bad time (2) ... but he found a new friend there, Steerforth. After David's mother died, Mr Murdstone sent him to work in a (3) ... in London. This was a terrible time.

Finally, David ran away to Dover to Aunt Betsey. She helped him and he went to a good school in Canterbury. He lived with Mr Wickfield and his (4) ..., Agnes. He also met (5) ..., an unpleasant, sinister man who worked for Mr Wickfield.

Later, David worked in London and fell in love with Dora Spenlow. More difficult times followed. Aunt Betsey (6) ... her money and Uriah Heep gained a lot of power over Mr Wickfield. Finally, David and Dora got married.

There was another problem for David. His friend Steerforth came from a rich family but he ran away with Emily, the niece of Mr Peggoty, an honest (7) ... . Steerforth didn't really love Emily and David helped Mr Peggoty to find her. Later, Steerforth (8) ... .

In Canterbury, Agnes was very unhappy. Uriah Heep wanted to marry her and her father was (9) ... Uriah. Luckily Mr Micawber, an old friend, was working for Uriah and knew his (10) ... . David was able to defeat Uriah Heep and Aunt Betsey got her money back. However, poor Dora died and David travelled abroad for (11) ... . When he returned, he married (12) ... who had always loved him. They were very happy.

0.	**A.** happy	**B.** sad	**C.** rich
1.	**A.** kind	**B.** cruel	**C.** friendly
2.	**A.** in Dover	**B.** in prison	**C.** at school
3.	**A.** market	**B.** factory	**C.** shop
4.	**A.** daughter	**B.** sister	**C.** wife
5.	**A.** Uriah Heep	**B.** Mr Micawber	**C.** Traddles
6.	**A.** increased	**B.** found	**C.** lost
7.	**A.** teacher	**B.** fisherman	**C.** lawyer
8.	**A.** went to Australia	**B.** died in bed	**C.** drowned
9.	**A.** friends with	**B.** afraid of	**C.** in control of
10.	**A.** secrets	**B.** mother	**C.** future plans
11.	**A.** five years	**B.** two years	**C.** six months
12.	**A.** Jip	**B.** Peggoty	**C.** Agnes

8 What do you think? Are these statements about the story true or false?

		True	False
0.	Aunt Betsey was a kind person.	T	☐
1.	David's mother was right to marry Mr Murdstone.	☐	☐
2.	Miss Murdstone was similar in character to her brother.	☐	☐
3.	Mr Peggoty and his family were poor but kind.	☐	☐
4.	Mr Murdstone chose a good school for David.	☐	☐
5.	It was easy for David to escape to Dover.	☐	☐
6.	David enjoyed his life in Canterbury.	☐	☐
7.	David's biggest mistake was to take Steerforth to Yarmouth.	☐	☐
8.	Mr Wickfield was the victim of Uriah Heep.	☐	☐
9.	Ham didn't love Emily.	☐	☐
10.	David Copperfield found true happiness with Agnes.	☐	☐

David Copperfield

KEY TO
THE EXERCISES
AND EXIT TEST

Charles Dickens and 'David Copperfield'

Page 11 – Exercise 1

1. C **2.** A **3.** A **4.** B **5.** C **6.** C **7.** A
8. B **9.** C **10.** B

Before you read

Page 14 – Exercise 1

1. d **2.** m **3.** g **4.** b **5.** h **6.** l **7.** e
8. k **9.** i **10.** c **11.** a **12.** f **13.** j

Part One

Page 19 – Exercise 1

1. B **2.** C **3.** A **4.** B **5.** A **6.** C **7.** C

Page 19 – Exercise 2

1. fisherman **2.** nephew **3.** uncle
4. hero **5.** servant

Page 20 – Exercise 3

If the spelling of an answer is not
correct, it will receive a mark if the word
is recognisable.

1. carrier **2.** people **3.** parcels
4. horse **5.** spend **6.** saved **7.** Peggoty
8. cooked **9.** cleaned **10.** message

Page 20 – Exercise 4

Open answer.

Part Two

Page 25 – Exercise 1

Possible answers:
1. Because he didn't answer the
   questions.
2. He bit Mr Murdstone's hand.
3. Peggoty.
4. 'Be careful - he bites.'
5. Traddles and Steerforth.
6. There was a new baby.
7. 'She's dead.'
8. To London.

Page 25 – Exercise 2

1. gave/forgot **2.** bit/hit **3.** sent/put
4. took/bought **5.** died/was
6. cried/heard

Page 26 – Exercises 3 & 4

Open answers.

Part Three

Page 31 – Exercise 1

1. money 2. Mrs Micawber
3. run away 4. only relative 5. bad
people 6. (small) cottage 7. excellent
8. believe

Page 31 – Exercise 2

1. H 2. D 3. B 4. F 5. E

Page 32 – Exercise 3

1. E 2. F 3. D 4. G 5. C 6. I 7. H
8. B
Open answers.

Page 33 – Exercise 4

1. Let's buy him some new clothes.
2. Let's wash him and put him to bed.
3. Let's fly my kite on the cliffs.
4. Let's call him 'Trotwood'.
5. Let's go to Canterbury.

Page 33 – Exercise 5

1. Rochester 2. Chatham 3. Faversham
4. Canterbury 5. Folkestone

Part Four

Page 38 – Exercise 1

Possible answers:
1. In Canterbury.
2. He was Mr Wickfield's clerk.
3. Doctor Strong.
4. Yes, he was.
5. Uriah began to have a strange power
over Mr Wickfield.
6. Mr and Mrs Micawber.
7. (At an inn) in London.
8. He wanted to meet the fishermen.

Page 38 – Exercise 2

1. D 2. F 3. B 4. H 5. G

Page 39 – Exercise 3

1. Uriah Heep is Mr Wickfield's clerk.
2. Doctor Strong's young wife is very
beautiful.
3. Mr Dick likes playing children's
games.
4. Doctor Strong has a boys' school.
5. The teachers' wages are high.
6. What's Mr Wickfield's daughter's
name? or What's the name of Mr
Wickfield's daughter?

Page 39 – Exercise 4

1. like a pig 2. like a swan 3. like a
monkey 4. like a lion 5. like a
hippopotamus 6. like a robot

Child Labour

Page 42 – Exercise 1

1. B 2. A 3. A 4. A 5. B 6. A 7. B
8. B

Page 42 – Exercise 2

Open answers.

Part Five

Page 46 – Exercise 1

1. A 2. A 3. B 4. B 5. C 6. A 7. C

Page 46 – Exercise 2

1. Doctors' Commons
2. Crupp
3. Highgate
4. 18 (years old)
5. 10.30

Page 47 – Exercise 3

1. B 2. A 3. C 4. A 5. B

Page 47 – Exercise 4

1. on 2. to 3. at 4. on 5. with
6. about 7. with 8. in 9. of 10. from

Part Six

Page 52 – Exercise 1

1. No, she wasn't.
2. Traddles and Mr and Mrs Micawber.
3. No, he wasn't.
4. A letter from Emily.
5. He destroyed the boat.
6. No, they weren't.
7. He decided to look for Emily in all the countries of Europe.
8. No, he wasn't.

Page 52 – Exercise 2

1. friend   2. James   3. mother   4. house
5. Oxford   6. (personal) servant
7. France, Germany

Page 53 – Exercise 3

1. lighthouse   2. nets   3. harbour
4. chips   5. horizon   6. castle   7. port
8. waves   9. ocean   10. crab

Part Seven

Page 58 – Exercise 1

1. picnic   2. goddess   3. jealous
4. rich, young woman   5. secret   6. lost
7. marry   8. aunts

Page 58 – Exercise 2

1. a diamond ring   2. a box of chocolates   3. a Valentine's card
4. a pearl necklace   5. a text message

Page 59 – Exercise 3

1. B   2. A   3. D   4. B   5. C   6. C   7. B
8. A

Page 59 – Exercise 4

Open answer.

Part Eight

Page 64 – Exercise 1

1. D   2. C   3. H   4. B   5. F   6. G   7. E

Page 64 – Exercise 2

1. 1839   2. 1841   3. Canada   4. Notes
5. Christmas Carol

Page 65 – Exercise 3

1. (The) County Hotel   2. (The) Thai House   3. Zizi's   4. (The) Fish Café
5. (The) Thruway Hotel

Page 65 – Exercise 4

1. in   2. a   3. and   4. wrote   5. there
6. in   7. would   8. to   9. Are
10. will/could/would/might

Dickens and Education

Page 68 – Exercise 1

1. C   2. B   3. B   4. B   5. A   6. C   7. B
8. C   9. B   10. A

Part Nine

Page 73 – Exercise 1

1. B   2. A   3. B   4. B   5. A   6. A   7. A

Page 73 – Exercise 2

1. C   2. B   3. C   4. C   5. B

Page 75 – Exercise 3

1. C   2. A   3. G   4. B   5. E

Page 75 – Exercise 4

Open answers.

Part Ten

Page 80 – Exercise 1

Possible answers:

1. Aunt Betsey told David not to criticise her.
2. He asked one of Emily's friends, Martha, to help him.
3. One day, the friend found David and took him to Martha's room.
4. Rosa Dartle was in the room.

5. He was happy.

6. He planned to emigrate to Australia with Emily.

Page 80 – Exercise 2

Here are some possible questions:

1. Where did Steerforth take Emily?

2. Who was Littimer?

3. Why did Martha help Mr Peggoty?

4. Where did they find Emily?

5. Why did Rosa Dartle shout at Emily?

6. What did Mr Peggoty do?

7. How did Ham feel?

Page 81 – Exercise 3

1. rescue **2.** search **3.** vanish
4. emigrate **5.** faint **6.** gossip **7.** steal
8. earn **9.** manage **10.** warn

(18) ### Page 82 – Exercise 4

1. Gad's Hill Place **2.** 1856 **3.** actress
4. separated **5.** (public) readings

Page 82 – Exercise 5

1. on **2.** It **3.** by **4.** well or easily
5. very **6.** is **7.** but **8.** from **9.** is
10. to

Part Eleven

Page 88 – Exercise 1

Possible answers:

1. A hypocrite, a serpent, a dishonest man, a liar.

2. Aunt Betsey and Traddles.

3. He stayed in the room.

4. Uriah's notebook.

5. In the fireplace.

6. Because he was afraid of prison.

7. She gave him some money.

8. Yes, they were.

(20) ### Page 88 – Exercise 2

1. June 9th **2.** 58 **3.** edited magazines
4. (John) Forster **5.** (Peter) Ackroyd

Page 89 – Exercise 3

1. C **2.** B **3.** A **4.** C **5.** B

Page 90 – Exercise 4

1. C **2.** B **3.** B **4.** D **5.** A **6.** A **7.** D
8. C

Page 90 – Exercise 5

Open answers.

Part Twelve

Page 96 – Exercise 1

1. A **2.** A **3.** A **4.** C **5.** B **6.** A **7.** C

Page 96 – Exercise 2

1. broke/ran **2.** blew/sank
3. Did/know **4.** was/died
5. travelled/thought **6.** forgot/began
7. knew/loved **8.** agreed

Page 98 – Exercise 1

1. A / 2. B / 3. B / 4. C / 5. A / 6. A /
7. B / 8. A / 9. C / 10. A

Page 98 – Exercise 2

1. Uriah Heep
2. Annie Strong
3. Jip
4. Barkis
5. Micawber

Page 99 – Exercise 3

1. Yarmouth
2. London
3. Dover
4. Canterbury
5. Commons
6. Australia

Page 99 – Exercise 4

1. of / 2. for / 3. to / 4. to / 5. of /
6. not / 7. marry / 8. with

Page 100 – Exercise 5

1. Peggoty
2. Mr Murdstone
3. Mr Creakle
4. Betsey Trotwood
5. Agnes
6. Dr Strong
7. Steerforth
8. Traddles
9. Mr Spenlow
10. Dora

Page 100 – Exercise 6

1. In 1812.   2. 'My favourite child'.
3. In a factory.   4. Because he owed
money.   5. London, Canterbury and
Yarmouth.   6. Mr Micawber.
7. Uriah Heep.   8. Queen Victoria.
9. No, it was rich and powerful.
10. The poor.

Page 100 – Exercise 7

1. B   2. C   3. B   4. A   5. A   6. C   7. B
8. C   9. B   10. A   11. B
12. C

Page 102 – Exercise 8

Probable answers:
1. F   2. T   3. T   4. F   5. F   6. T
7. T   8. T   9. F   10. T

Notes

Notes

Notes

Black Cat English Readers

BLACK CAT ENGLISH CLUB

Membership Application Form

BLACK CAT ENGLISH CLUB is for those who love English reading and seek for better English to share and learn with fun together.

Benefits offered: - *Membership Card*

- *Member badge, poster, bookmark*

- *Book discount coupon*

- *Black Cat English Reward Scheme*

- *English learning e-forum*

- *Surprise gift and more...*

Simply fill out the application form below and fax it back to 2565 1113.

Join Now! It's FREE exclusively for readers who have purchased *Black Cat English Readers* !

The book(or book set) that you have purchased: _____

English Name:_____ (Surname) _____ (Given Name)

Chinese Name: _____

Address:_____

Tel: _____ Fax: _____

Email:_____
Sex: ❏ Male   ❏ Female   (Login password for e-forum will be sent to this email address.)

Education Background: ❏ Primary 1-3   ❏ Primary 4-6   ❏ Junior Secondary Education (F1-3)

❏ Senior Secondary Education (F4-5)   ❏ Matriculation

❏ College   ❏ University or above

Age: ❏ 6 - 9   ❏ 10 - 12   ❏ 13 - 15   ❏ 16 - 18   ❏ 19 - 24   ❏ 25 - 34

❏ 35 - 44   ❏ 45 - 54   ❏ 55 or above

Occupation: ❏ Student   ❏ Teacher   ❏ White Collar   ❏ Blue Collar

❏ Professional   ❏ Manager   ❏ Business Owner   ❏ Housewife

❏ Others (please specify: _____)

As a member, what would you like **BLACK CAT ENGLISH CLUB** to offer:

❏ Member gathering/ party.   ❏ English class with native teacher   ❏ English competition

❏ Newsletter   ❏ Online sharing   ❏ Book fair

❏ Book discount   ❏ Others (please specify: _____)

Other suggestions to **BLACK CAT ENGLISH CLUB**:

_____

Please sign here: _____

(Date: _____)